METAMORPHOSIS

STAGES IN A LIFE

David Suzuki

Stoddart

First published in 1987 by
Stoddart Publishing Co. Limited
34 Lesmill Road
Toronto, Canada
M3B 2T6

Second printing October 1987

Canadian Cataloguing in Publication Data

Suzuki, David T., 1936-
Metamorphosis: stages in a life

ISBN 0-7737-2139-8

1. Suzuki, David T., 1936- . 2. Geneticists - Canada - Biography. 3. Broadcasters - Canada - Biography. I. Title.

QH429.2.S89A3 1987 5757.1'092'4 C87-094241-7

Editor: Sarah Swartz, The Editorial Centre
Text illustrations: Rob Tuckerman
Jay Tee Graphics Ltd.
Jacket photograph: James Murray
Jacket design: Falcom Design & Communications
Typeset by Jay Tee Graphics Ltd.
Printed in the United States

TABLE OF CONTENTS

To My Father

This book has become reality because of my father, Kaoru Carr Suzuki. From the genes you gave me, to the lifetime of experience and example, you helped me through much of my metamorphosis. Thanks, Dad.

PREFACE

The question that I am asked most often is, "Why did you go into television?" and the second most frequent query is, "What got you into science?" I have two answers: "Pearl Harbor" for the first, and "Dad" for the second. This book is an attempt to flesh out my cryptic replies.

For me, the answers to these questions are inextricably bound up in the events that have shaped me from childhood on. The answers, therefore, can only be understood within the context of my life. I know that my life has not been more special or important than that of any other person, but while science is the major force shaping society today, scientists remain shielded from the public by a veil of jargon and expertise. Who are they? What do they do? What drives them on? And within the community of scientists, the number who take up broadcasting and become celebrities is minuscule. Perhaps that makes my life instructive and this book worth reading.

There is no such thing as objective reality. We each select our experiences through the filters of our genes, values and belief systems. Ask a logger from British Columbia and a Haida Indian to talk about the importance of a virgin stand of cedar trees, or a Bay Street businessman and a person on welfare about taxes and social services. You would swear from their responses that they come from different worlds. Beginning with the world view of our parents and then that of our surrounding society and institutions, we learn to see the world in a limited way that influences what we remember or forget, emphasize or downplay. Essentially, this book is a similar construction done in a more obvious and deliberate way. By writing this book I have made up a story of my life. Were my parents, first wife, best friend, student or bitter opponent to write about me, the stories would be different.

I have recounted experiences, events and memories that seemed important to me and which may provide insight into some of the ideas and opinions I now hold. The person you see emanating from a television tube or hear from the speaker of a radio is a creation of the technology. Perhaps this book will add some meat to that electronic persona. But my real hope is not so much to introduce myself as to provide a different angle from which the reader may see her or his own experiences, values and beliefs.

The organizing principle in this book is a rough chronology, beginning with my ancestors and proceeding through my childhood, adolescence and maturity. But in order to complete some topics, for example a discussion of each of my parents, chronology is bent so that I can bring in events that are covered in more detail later.

I have chosen to identify by name many people whom I have admired or to whom I have been grateful. I want to acknowledge them. Those who I recall in a negative way are not named, although their descriptions, times and events are presented in sufficient detail to allow anyone who cares to identify them. I've omitted names, not from fear of litigation, but because there is no point. I mention certain incidents merely to illustrate how they affected me and my life at that time.

In retrospect, my life has been marked by a series of transformations. It's interesting to note that in the rest of the biological world, profound change in the lives of many organisms is a natural and necessary part of their development. Often these changes involve dramatic transitions in physical makeup, behaviour and habitat. This process is called *metamorphosis.*

Fruitflies have been a passion in my life for three decades. In reflecting on their heredity, behaviour and life cycle, I've come to see that in many ways, the changes in our lives parallel those remarkable stages that take place in a fly's life. All life forms receive a genetic legacy from their ancestors in the form of DNA, the chemical blueprint that shapes the way we are. Each fly originates within the protective membrane of an egg just as we began our lives in the cozy security of a womb. The fly egg releases a larva which is a sophisticated organism in its own right, before it becomes a fruitfly. And similarly, from the womb emerges a newborn baby bearing no resemblance to the fertilized egg, not yet a child nor an adult.

Both baby and larva possess the capacity to respond to their environments, albeit in a limited fashion. As they grow larger, they increase the repertoire of their behaviour as they learn and respond to environmental cues. They become increasingly mobile; they are able to search on their own. A fly undergoes successive moults to become second- and third-stage larvae. A child reaches stages in maturation: walking, talking, puberty and adolescence.

At a certain point, cued by size, temperature, hormones, season or other factors, many insects undergo a dramatic change from larva to adult through a process called pupation. Within the pupal case, the nervous system of the larva is retained, but the larval carcass is broken down to be digested and reorganized. Embryonic clusters of cells predestined to form parts of the adult then spring into action and grow and form a new creature. Emerging from the pupal case is a dramatically different organism with eyes, wings, legs and gonads, that will take it into dimensions far beyond the limited scope of its previous stage. What cues do we as human beings respond to and what do we become after the transformation?

In both child and immature fly, each stage is impressively larger than the preceding one but still incomplete. An adult person has to digest accumulated experiences, information and knowledge in order to leave the immature state comparable to a fly's maggot-hood. Many people remain maggots, growing larger, richer and more powerful without an accompanying evolution in wisdom, sensitivity or compassion.

This metaphor is far from perfect, but in looking back on my life, I can now recognize changes that signal the stages of my metamorphosis. I have yet to determine whether I've left maggot-hood behind me.

I have, like most people, done a few things in my life of which I am proud and many that I regret. In making important decisions, where I've had the luxury of prior reflection, I've tried to guide my actions by the answer to the questions: If I do or do not do this, will my parents still be proud of me, will my wife still respect me, and will my children continue to love me? If the answer is "yes"to all three, then the way is clear. That's what I did before I decided to act on my conceit and write this book.

David Suzuki
Toronto, 1987

ACKNOWLEDGEMENTS

I would first like to thank Ed Carson who, while working with Stoddart Publishing, persuaded me to consider writing this book. My immediate reaction to his suggestion was that my life is of little interest and I am too young. I had really wanted to publish a collection of essays I'd written over the years, but Ed convinced me that autobiographical material was essential to give the essays broader interest. The reader will judge who was right, but I am grateful for the exercise of collecting together and reflecting on the records of my past. It was incredibly self-indulgent, unexpectedly enjoyable, frequently excruciating, but ultimately informative, revealing and cathartic for me. I am grateful to Ed for that.

This book is as much a creation of my editor, Sarah Swartz, as it is mine. She took the raw manuscript and gave it shape, pulled stuff out of me I didn't know was there, deleted a lot of extraneous material and yet left it in my voice. Thank you for all of it, Sarah; you will never get all of the credit you deserve. But I know.

Thanks, Dad, for spending so many hours going over the past in such detail and for verifying and correcting my version of events.

Shirley Macaulay was her usual indispensable self, bringing

me down to earth when I was flying too high, commenting on the manuscript in ways that led me to revise it, and producing drafts as fast as I got material to her. You are not allowed to retire before I clone you, Shirley.

And I have to thank my wife, Tara Cullis. First and most important, she gave me her opinions of my work that were by far the most valued, professionally and personally. Often the manuscript was as painful for her to read as it was for me to write. She reviewed it with love and respect without losing her critical facility. She encouraged me during my low points and exulted with me when I was high. And most important, she never complained as I left her with a disproportionate amount of the housework and the children to look after while I was banging this out. Thank you, Tara; I'll do the same for you when you do your magnum opus.

I relied heavily on Ken Adachi's excellent book, *The Enemy that Never Was*, for the history of the Japanese in Canada and the evacuation.

Writer John Blazina reviewed the very first few pages of this book, gave me useful suggestions, and didn't suggest I should forget all about it.

And finally, many different people at Stoddart Publishing, from Jack Stoddart on, have encouraged me and offered their generous help and support to make this project a happy experience.

I was also aided by the Public Awareness Program for Science and Technology, Serial GP-86-00012.

PHOTO CREDITS

1, 2, 3, 4/ Roy Kumano, London, Ontario; 5/ CBC Radio; 6/ John Crawford; 7/ CBC TV; 8/ National Film Board of Canada; 9/ Terry McCartney-Filgate, National Film Board of Canada 10/ Don Kishibe; 11, 12, 13, 14, 15, 16/ CBC TV.

PROLOGUE

My grandparents emigrated from Japan to this country at the turn of the century. Like so many immigrants, they left their homeland reluctantly. But they came from a poverty so profound that they were prepared to take the risk and deal with the terrifying unknown of a totally alien culture and language. Their children, my parents, were born in Vancouver over seventy-five years ago. They were Canadians by birth. By culture, they were genuine hybrids, fluently bilingual but fiercely loyal to the only country they had ever known — Canada.

On December 7, 1941, an event took place that had nothing to do with me or my family and yet which had devastating consequences for all of us — Japan bombed Pearl Harbor in a surprise attack. With that event began one of the shoddiest chapters in the tortuous history of democracy in North America. More than twenty thousand people, mostly Canadians by birth, were uprooted, their tenuous foothold on the West Coast destroyed, and their lives shattered to an extent still far from fully assessed. Their only crime was the possession of a common genetic heritage with the enemy.

Although I have little recollection of that time, Pearl Harbor was the single most important event shaping my life; years later in reassessing my life during a personal trauma, I realized that virtually every one of my emotional problems went right back to it.

Throughout the entire ordeal of those war years, my parents acted with a dignity, courage and loyalty that this young country did not deserve. Today, my mother is dead, never having known the symbolic acknowledgement that a wrong was committed against her. But if there is anything worthwhile to be salvaged from those years, it is that her story and my father's, through me, will not be forgotten and will serve as a legacy to all Canadians, a reminder of the difficulty of living up to the ideals of democracy. The stories of how my parents and their parents fared in Canada are both a tribute to their strength of character and a record of the enormous changes that have occurred in this country. Whatever I am has been profoundly shaped by these two facts.

CHAPTER ONE

ANCESTORS — THE GENETIC SOURCE

MY GENES can be traced in a direct line to Japan. I am a pure-blooded member of the Japanese race. And whenever I go there, I am always astonished to see the power of that biological connection. In subways in Tokyo, I catch familiar glimpses of the eyes, hairline or smile of my Japanese relatives. Yet when those same people open their mouths to communicate, the vast cultural gulf that separates them from me becomes obvious: English is my language, Shakespeare is my literature, British history is what I learned and Beethoven is my music.

For those who believe that in people, just as in animals, genes are the primary determinant of behaviour, a look at second- and third-generation immigrants to Canada gives powerful evidence to the contrary. The overriding influence is environmental. We make a great mistake by associating the inheritance of physical characteristics with far more complex traits of human personality and behaviour.

Each time I visit Japan, I am reminded of how Canadian I am and how little the racial connection matters. I first visited Japan in 1968 to attend the International Congress of Genetics in Tokyo. For the first time in my life, I was surrounded by people who all looked like me. While sitting in a train and looking at the reflections in the window, I found that it was hard to pick out my own image in the crowd. I had grown up in a Caucasian society in which I was a minority member. My whole sense of self had developed with that perspective of looking different. All my life I had wanted large eyes and brown hair so I could be like everyone else. Yet on that train, where I did fit in, I didn't like it.

On this first visit to Japan I had asked my grandparents to contact relatives and let them know I was coming. I was the first in the Suzuki clan in Canada to visit them. The closest relative on my father's side was my grandmother's younger brother, and we arranged to meet in a seaside resort near his home. He came to my hotel room with two of his daughters. None of them spoke any English, while my Japanese was so primitive as to be useless. In typical Japanese fashion, they showered me with gifts, the most important being a package of what looked like wood carved in the shape of bananas! I had no idea what it was. (Later I learned the package contained dried tuna fish from which slivers are shaved off to flavour soup. This is considered a highly prized gift.) We sat in stiff silence and embarrassment, each of us struggling to dredge up a common word or two to break the quiet. It was excruciating! My great uncle later wrote my grandmother to tell her how painful it had been to sit with her grandson and yet be unable to communicate a word.

To people in Japan, all non-Japanese — black, white or yellow — are *gaijin* or foreigners. While *gaijin* is not derogatory, I find that its use is harsh because I sense doors clanging shut on me when I'm called one. The Japanese do have a hell of a time with me because I look like them and can say in perfect Japanese, "I'm a foreigner and I can't speak Japanese." Their reactions are usually complete incomprehension followed by a sputtering, "What do you mean? You're speaking Japanese." And finally a pejorative, "Oh, a *gaijin*!"

Once when my wife, Tara, who is English, and I went to Japan, we asked a man at the travel bureau at the airport to book a *ryokan* — a traditional Japanese inn — for us in Tokyo. He found one and booked it for *"Suzuki-san"* and off we went. When we arrived at the inn and I entered the foyer, the owner was confused by my terrible Japanese. When Tara entered, the shock was obvious in his face. Because of my name, they had expected a "real" Japanese. Instead, I was a *gaijin* and the owner told us he wouldn't take us. I was furious and we stomped off to a phone booth where I called the agent at the airport. He was astonished and came all the way into town to plead our case with the innkeeper. But the innkeeper stood firm and denied us a room. Apparently he had accepted *gaijin* in the past with terrible consequences.

As an example of the problem, Japanese always take their shoes off when entering a *ryokan* because the straw mats *(tatami)* are quickly frayed. To a Japanese, clomping into a room with shoes on would be comparable to someone entering our homes and spitting on the floor. Similarly, the *ofuro,* or traditional tub, has hot clean water that all bathers use. So one must first enter the bathroom, wash carefully and rinse off *before* entering the tub. Time in the *ofuro* is for relaxing and soaking. Again, Westerners who lather up in the tub are committing a terrible desecration.

To many Canadians today, the word "Jap" seems like a natural abbreviation for Japanese. Certainly for newspaper headlines it would seem to make sense. So people are often shocked to see me bristle when they have used the word Jap innocently. To Japanese-Canadians, Jap or Nip (from "*Nippon*") were epithets used generously during the pre-war and war years. They conjure up all of the hatred and bigotry of those times. While a person using the term today may be unaware of its past use, every Japanese-Canadian remembers.

The thin thread of Japanese culture that does link me to Japan was spun out of the poverty and desperation of my ancestors. My grandparents came to a Canadian province openly hostile to their strange appearance and different ways. There were severe restrictions on how much and where they could buy property. Their children, who were born and raised in Canada, couldn't vote until 1948 and encountered many barriers to professional training and property ownership. Asians, regardless of birthplace, were third-class citizens. That is the reality of the Japanese-Canadian experience and the historical cultural legacy that came down to the third and fourth generations — to me and my children.

The first Japanese immigrants came to Canada to make their fortunes so they could return to Japan as people of wealth. The vast majority was uneducated and impoverished. But in the century spanning my grandparents' births and the present, Japan has leapt from an agrarian society to a technological and economic giant.

Now, the Japanese I meet in Japan or as recent immigrants to Canada come with far different cultural roots. Present-day Japanese are highly educated, upper-middle class and proud of their heritage. In Canada they encounter respect, envy and curiosity in sharp contrast to the hostility and bigotry met by my grandparents.

Japanese immigrants to North America have names that signify the number of generations in the new land (or just as significantly, that count the generational distance *away* from Japan). My grandparents are *Issei,* meaning the first generation in Canada. Most *Issei* never learned more than a rudimentary knowledge of English. *Nisei,* like my parents, are the second generation here and the first native-born group. While growing up they first spoke Japanese in the home and then learned English from playmates and teachers. Before the Second World War, many *Issei* sent their children to be educated in Japan. When they returned to Canada, they were called *Kika-nisei* (or *Kibei* in the United States). Most have remained bilingual, but many of the younger *Nisei* now speak Japanese with difficulty because English is their native tongue. My sisters and I are *Sansei* (third generation); our children are *Yonsei.* These generations, and especially *Yonsei*, are growing up in homes where English is the only spoken language, so they are far more likely to speak school-taught French as their second language than Japanese.

Most *Sansei,* like me, do not speak Japanese. To us, the *Issei* are mysteries. They came from a cultural tradition that is a hundred years old. Unlike people in present-day Japan, the *Issei* clung tightly to the culture they remembered and froze that culture into a static museum piece like a relic of the past. Not being able to speak each other's language, *Issei* and *Sansei* were cut off from each other. My parents dutifully visited my grandparents and we children would be trotted out to be lectured at or displayed. These visits were excruciating, because we children didn't understand the old culture, and didn't have the slightest interest — we were Canadians.

My father's mother died in 1978 at the age of ninety-one. She was the last of the *Issei* in our family. The final months of her life, after a left-hemisphere stroke, were spent in that terrible twilight — crippled, still aware, but unable to communicate. She lived the terminal months of her life, comprehending but mute, in a ward with Caucasian strangers. For over thirty years I had listened to her psychologically blackmailing my father by warning him of her imminent death. Yet in the end, she hung on long after there was reason to. When she died, I was astonished at my own reaction, a great sense of sadness and regret at the cleavage of my last link with the source of my genes. I had never been able to ask what made her and others of her generation come to Canada, what they felt when they arrived, what their hopes and dreams had been, and whether it was worth it. And I wanted to thank her, to show her that I was grateful that, through them, I was born a Canadian.

SOME JAPANESE BACKGROUND

Japan's population is clumped around ports and the parts of the islands that are arable. Japan has five times as many people as Canada, with a total area less than the size of our Maritime provinces. And most of the land is mountainous and uninhabitable. Through centuries, perhaps millenia, pockets of Japanese people developed in relative isolation. Each region was called a *prefecture,* a political delineation like a province or state. Like Greek city states, each group evolved with distinctive social, economic and linguistic characteristics that are still recognizable.

The local insularity also applied externally. Japanese people were protected by the seas from "contamination" by other cultures. For centuries they deliberately resisted influences from the rest of the world. Finally, in the mid-sixteenth century, Portuguese sailors were blown onto islands near Kyushu, and were the first Europeans to set foot on Japan. They brought with them guns and Christianity, both of which had an explosive impact on Japan.

The muskets brandished by those sailors were carefully duplicated by the Japanese and enabled Oda Nobunaga, a lord of central Japan, to win a number of critical battles. In 1603, his dream of a unified country was realized when Iyeyasu established himself as a *shogun* (military dictator). He was the first ruler of the Tokugawa era which ended the struggles between feudal lords and established law and order for two-and-a-half centuries.

With this era began yet another period of deliberate, enforced isolation from the rest of the world. Worried about the rapid spread of Christianity, the Tokugawas persecuted Christians relentlessly. For the next two centuries, Japan wrapped herself in an impenetrable cocoon while Europe exploded with the vitality of the Renaissance. Peace and order in isolated Japan reinforced a sense of community that remains a major feature of Japanese culture.

The isolation came to an end on July 8, 1853 when steam-driven ships, the like of which they had never seen before, under the command of the American Matthew Perry anchored in Yedo Bay. In the face of the technological superiority of this force, Japan was helpless. On March 31, 1854, the Treaty of Kanagawa was signed and Japan was forced to open her ports to foreign commerce.

The Tokugawa era soon came to an end as a rebel movement to overthrow the Shogunate and restore the emperor gained strength. In 1867, rebels launched a successful coup, ousting the Tokugawas and bringing a fifteen-year-old boy to head the new era. Thus, in the year of Canada's birth, Japan also began a new life. The young ruler became known as the Emperor Meiji. He presided over an unprecedented transition from feudalism to a modern industrial state, remarkable for occurring almost overnight.

Japan now rushed to make up for the lost time, and in the tradition of learning by imitation Japanese people aped Western clothes and customs. The Japanese are not ashamed of their ability to imitate. They have a saying, "The best things come in boats." They acknowledge that they have borrowed the best from India, China and, most recently, America.

During the Tokugawa rule, social classes and standing had been rigidly observed. Below the rulers were the warriors, followed in descending order by peasants, artisans and merchants. Only the *samurai* or warrior class was allowed to bear arms. Magnificently skilled in the martial arts, they were the instruments of enforcement of law and order. In return, they were rewarded handsomely. They became wealthy landowners, aristocracy in their own right. In 1874, compulsory military service was imposed on peasants and with that the end of the *samurai* was in sight.

The first decade of the Meiji era was not without tremendous social unrest and upheaval. The transition was explosive, and during many agrarian revolts the *samurai* were able to sell their services as *ronin,* wandering fighters for hire. But their skills in the new social order were outmoded; modernization had rendered them irrelevant.

Although peasants ranked right below *samurai* in the social structure, they were squeezed mercilessly by their landlords. With modernization, the plight of the peasant grew worse. While these peasants were still at the base of the economy, the new leaders were preoccupied with other things as they rushed to industrialize the country.

It was the Meiji Reformation that eventually led to Japanese emigration to distant lands to escape the shackles of poverty. That migration made possible in Canada what would never have happened in Japan: the wedding of my mother and father — descendants from different prefectures and very different classes.

MY FAMILY: THE NAKAMURAS

My mother's parents were unusual amongst the first wave of immigrants from Japan; they were relatively high born and educated. My mother's mother was from the prefecture of Hiroshima. She and her husband came to Canada independently and married through an arrangement. I only knew my maternal grandparents until the beginning of World War Two, as they returned to Japan shortly after the end of the war.

The daughter of a Buddhist priest, my grandmother remained a practising and devout Buddhist all her life. My mother was raised as a Buddhist, and although she converted to Christianity, I always felt she had never really changed her religious outlook. My grandmother was the central figure in mom's life. My mother told me at my grandmother's death that she was sure she would be with her again, even though she didn't have a Christian sense of heaven.

My grandmother had been trained to be a nurse when she came to Canada. During the epidemic of Spanish flu, people were isolated in temporary quarters set up in schools. They died like flies, while grandmother tended to them and escaped without getting sick. I have met people who referred to her as an angel.

Grandfather Nakamura was from the Kumamoto prefecture and came from a *samurai* background. His father was one of the victims of the Meiji Restoration, but not before his son had been raised to expect better things. Grandfather had been taught the martial arts and horseback riding. He claimed, for example, there was a way of drawing the testicles up into the body to protect them when going into battle.

Once he was asked to round up a group of Japanese men to do some logging. It was only when their train pulled in to the station that he discovered that his men were "scabs" brought in to break a strike of loggers. When the men disembarked, they found a wall of strikers waiting to stop them. The Japanese men were trapped and terrified. My grandfather realized there was nothing to do but brazen it out and he told the men to follow him. His carriage and demeanour reflected his *samurai* background and he led the men straight up to the loggers, who wavered and stepped aside to let them through.

Physical labour was beneath my grandfather's station, so rather than struggling for survival in Japan he chose to seek his fortune in North America. He hoped to return to Japan a wealthy man. But life didn't turn out as he expected. While Japan has a long and rich history, with a culture that is complex and sophisticated, British Columbia at the turn of the century was rugged and uncivilized — a frontier with few traditions and little culture. Grandfather tried a number of get-rich-quick schemes, such as exporting salted herring and shark to Japan, but they never succeeded. He never worked at manual labour and acted as he had been raised, as an aristocrat. My memories of him are few. To me, he seemed an immense, imposing figure who indulged us by buying whatever we wanted. He was like an innocent in the business world, just waiting to be taken advantage of — and he often was. His presence commanded attention, but it was his wife who worked, brought in the money and managed the household.

In Japanese society, males are the favoured sex. They are the masters of their households; their word is law. Men most often lead social and business transactions, while women and children offer their silent, dutiful presence. This is not a caricature; it is a portrait of both sides of my family. My experience is also that Japanese women are the rock-solid foundations of the family. While men go about performing publicly, it is women who raise children, manage accounts, cook and clean, and often work outside the home to supplement incomes.

Like many immigrants to Canada, my maternal grandparents had no intention of staying forever. They had four children, two sons and two daughters. The baby of the family was my mother, Setsu. Those children were Canadians by birth. They were also registered by their parents as Japanese nationals. And this is what led to the oft-repeated claims that the Japanese citizenship of these

Canadian-born people showed where their loyalties lay. In fact, that second generation, the *Nisei*, had not chosen loyalties; it was their parents who had registered them as Japanese citizens in anticipation of returning to Japan. The *Nisei* were *Canadians* by both birth and culture. They should have been advantaged by straddling two great cultural traditions, but in the thirties, this was a trap. Never would they find total acceptance as Canadians, yet they most definitely were not Japanese.

This was tragically true in the case of my mother's second brother. He was Satoshi, nicknamed Sally. He and his older brother, Frank, had played for the Asahis, a popular baseball team in Little Tokyo. A skilful athlete, Sally was also a good singer and a real ham. So in the mid-'30s, he left for Japan to seek his fortune as a singer. While living in Tokyo, he met and married a Japanese. However, once Japan attacked Manchuria, he was trapped in Japan. He continued to work in show business and eventually worked in the same place as Iva Toguri. She was the infamous Tokyo Rose, an American-born Japanese who broadcast programs to Allied troops, urging them to surrender. After the war, when Toguri was brought to trial in the United States for treason, Uncle Sally was brought over to testify. We were able to see him for the first time in a decade.

Uncle Sally and his wife had three children and he carved out a successful career in Japan as an actor and television performer. He recorded a hit single, a Japanese version of "High Noon," and became known as Japan's Tex Ritter. Once, on a trip to Cologne, West Germany, I went to see an American western movie and was amazed to find Uncle Sally playing a misplaced *samurai* in the American West.

But Sally had always felt trapped in Japan, and his greatest dream was to return "home" to Canada. He had done the reverse of his father. He had never given up his Canadian citizenship and had registered each of his three Japan-born children as Canadian nationals. Finally, in 1969, he simply pulled up stakes — sold his house, quit his job, broke off his daughter's arranged marriage, pulled his son out of Tokyo University — and moved his family to Vancouver. During all those years away, Sally had nursed the memory of Canada as it had been when he left in 1937! He returned to a Vancouver from which the Japanese had been uprooted and dispersed. Little Tokyo was gone and the people were spread across the country. It was a Rip Van Winkle story. When I tell this story to Japanese people, they immediately say *Urashimataro,* the title of a similar Japanese folk tale.

In the ancient Japanese story, Urashimataro is a boy who rescues a turtle from some other boys and returns it to the sea. One day, a turtle comes back and takes the boy into the sea with him. There Urashimataro finds a marvelous underwater city where he is clothed in the finest material and dines on the most elegant food. He leads a good easy life. But one day he becomes homesick and wishes to see his mother and father. He begs to be returned to his village. His wish is granted and he is taken back only to find that while he has not aged, time has passed more quickly on land. Everyone he knew is now dead and he is a stranger. So Urashimataro became a person who belonged neither in the city under the sea nor in his original village. This was also the case for Uncle Sally.

I saw a lot of Sally when he arrived in Vancouver. I helped him get settled and look for a job. But his family was devastated. They had come from Tokyo, the centre of their universe. They hadn't come here in desperation and poverty as my grandparents had. Compared to Tokyo, Vancouver was dull and uninteresting and everyone spoke an alien language. They were homesick and unhappy. One night when I dropped in, my uncle broke down and wept. "I have come back to find I am a foreigner. I'm not a Canadian any more, I am Japanese." My uncle Frank, who was living in Toronto, had told him not to return to Canada. But Sally had defied him and had come on his own. And now he couldn't stand the loss of face of having to admit that he had been wrong. There was I, barely thirty, telling him to forget his brother, take responsibility for his family first and "go home." He hugged me and cried in gratitude, and soon after he and his family went back to Japan.

In spite of his family's misery, Uncle Sally had almost stayed in Canada out of necessity to "save face." There is no question that "face" is very important to a Japanese. My father never forgot nor forgave what he perceived as insults or slights by others. Often he imagined them or blew them up out of proportion. I've seen him cut all ties with someone he was once very friendly with, just because he thought the person had been insulting or rude. Where dad will go out of his way to help someone out or to give a special gift, he's quite content with a "thank you" phone call or letter. But if that acknowledgement is overlooked, dad may never speak to that person again.

Even as a third-generation Canadian, I have a strong need to preserve face. To me, Caucasians have a wonderful capacity of saying, "I'm sorry" or accepting an apology and then forgetting the incident, letting bygones be bygones. Like dad, I find myself being hurt or angered at a student's flippant remark, an associate's offhand crack or my wife's insouciance, and can't just let it drop.

Sometimes this has been tough for my wife, Tara. In a Japanese family, wives always support their husbands in public. I love the kind of give and take in Caucasian families where wives express their opinions openly and forcefully. And Tara does this well. But on occasion when I do or say something really stupid and in front of others, she won't let me get away with it. She crosses that line and I end up exploding or glowering in a sulk. I've lost face. It's a terrible trait; it's too often petty and self-indulgent. In Japan, it may impose a certain standard of behaviour that maintains social stability, but in Canada, it's a pain.

By a fluke of war, Uncle Sally had been trapped in Japan at the same time that Japanese-Canadians were trapped by racism and fear in Canada. The vast majority of Japanese-Canadians regarded British Columbia as home, but after Pearl Harbor, their civil rights were suspended, their property was confiscated and they were incarcerated in camps in British Columbia's interior. It was a devastating blow — economically, socially, psychologically. And as the Second World War drew to a close, Japanese-Canadians were faced with yet another indignity — the mandatory choice to sign up to go to Japan or leave the West Coast and disperse east of the Rockies. Many Japanese-Canadians, encouraged by the government of British Columbia, decided to go to Japan.

My mother's parents signed up to "repatriate," as did my mother's sister and her husband (who was an *Issei).* They left for Japan shortly after the war, while we obeyed the government's orders and dispersed out east to Ontario. Within a year, both of my mother's parents were dead, my grandmother from a severe stroke and my grandfather from senility and malnutrition. But my parents always believed the real cause of their deaths was simply culture shock and disillusionment.

After the war, letters to and from Japan took at least a month to get to their destination. By the spring of 1946, we were living on a farm in a rural part of southern Ontario. One night, my mother woke up crying and told my father her mother had died. A month later, she received a letter from her sister telling her that this had indeed happened.

To my regret, I last saw my maternal grandparents when I was about seven. Through my child's eyes, they had a dignity that came from a higher social class in striking contrast to the earthy style of my father's parents. My mother's parents are only a vague memory for me. But because mom often talked about them, I knew she would be very happy if I ever paid a visit to the temple where their ashes were kept. I got a chance in 1972 after Tara and I were married. We spent four months travelling around the world, finishing with a month in Japan. It was my second visit.

We decided to make a pilgrimage to the Buddhist shrine where my maternal grandparents' ashes were kept. We left Hiroshima and took a number of local trains and finally a taxi to the temple. There we met the son of the priest who ran the temple, and with my primitive Japanese I was able to identify myself. We were amazed to discover a distant relative still living at the temple. She was a ninety-two-year-old woman, who was mom's aunt, all bent and gnarled from years of work and kneeling. She told us one of my mother's brothers was supposed to be trained to be a priest in the temple. But since neither of them was willing, an outside family was brought in to substitute. As a Nakamura, I was treated with great respect. A big fuss was made over our visit and, after an elaborate meal, we were treated to a special ceremony laid on just for us.

The young man was training as a priest and donned all of the elaborate paraphernalia of the priesthood. He ushered us into a huge temple with an ornate altar and burning incense. He struck a gong, chanted a long incantation, burned more incense and then indicated that on the altar were the remains of my grandparents.

It was a moving ceremony, and we were overwhelmed by emotion by the time he left us alone as we kneeled in front of the altar. After several minutes of silence, we decided to take a closer look at the altar and crawled over to find the ashes. Featured on the ledge like icons were two tin cans with paper labels that identified their contents as mandarin oranges. After a few minutes of silence, I mumbled about the ignominious treatment of my grandparents — to end up stuffed into two mandarin orange cans. Tara burst into tears. It took us half an hour to get up the nerve to pick up the tins — and to find they were, in fact, unopened and meant as offerings to the spirits of the dead. We heard later that the ashes of my grandparents were contained in elegant lacquer boxes somewhere *inside* the altar.

MY FAMILY: THE SUZUKI LINE

In contrast to my mother's parents, dad's mother and father were uneducated and of the peasant class. They both came from the prefecture of Aichi and had married by arrangement in Japan. My grandmother's parents were farmers, and she used her farming experience when she came to Canada by raising chickens and growing vegetables for sale. My grandfather's father died when he was twelve years old, and he was immediately apprenticed out to be a carpenter. He was so skilled that when he became a boatbuilder in Canada, he measured planks by eye. It is reputed that when his planks were cut, they fitted perfectly every time.

My grandparents came to Canada before World War I as *yobi yose* or "call-ups." They were sponsored by friends who had emigrated earlier from their village. In return, they promised that they would work for the sponsor for a couple of years. They came to a hostile province gripped with the fear of the "Yellow Peril." British Columbia was as dazzlingly beautiful and rich in resources as my grandparents had been led to believe. But they did not expect the hostility and the socio-economic barriers. They were forced, mainly by the chasm of language, to rely on the few Japanese who could speak English and who were savvy to the strange ways of Canadians. My grandparents often were totally dependent on Japanese entrepreneurs who knew their way around.

Grandfather was diligent and quick to learn. His first job was to work as a salmon fisherman and he learned it well. After he completed his service, he added a job as a boatbuilder. The Suzuki family had seven children, six boys and a girl. His sons became boatbuilders and fishermen, too.

The family eventually settled in Marpole, a village at the south end of Vancouver on the edge of the Fraser River. Back then, it was permissible to raise livestock on one's property within city limits, and my grandfather raised chickens for eggs and meat. My father was the eldest in the family. He and the next son, Mar, had the chore of cleaning the chicken houses of manure, packaging it and selling it to farmers for fertilizer.

Like my mother's father, and all who live simply but need money desperately, my father's father was gullible and easily persuaded to invest in people's schemes to make money. With the family's carefully saved money, he would make loans and watch them disappear. But there were always Japanese people with surefire schemes to get rich, and grandfather never learned. As a result of this, my father admonished me never to lend money if I valued friends. "If you want to help a person out, then *give* him the money," dad advised, "because otherwise, if he can't repay you, it will eat you up inside and you'll lose a friend. That's what it did to my father." Grandfather never struck it rich, but it wasn't for lack of trying.

It was the custom for Japanese emigrants to send their children to Japan to get their education and relieve the family of rearing the youngsters in their early years. When my father was five years old, he and his two younger brothers were sent to Japan with his mother. He had overheard his parents talking and knew that he and his brothers were to stay behind in Japan with his mother's parents. He didn't want to stay in Japan, so he developed a strategy. Whenever he was left with his relatives, and his mother wasn't in sight, he started to howl and kick up a fuss. He got to be such a pain in the neck that after a month of this his grandmother insisted that the three boys be taken back to Canada. That month in Japan was burned indelibly in my father's memory; when I took him back to Japan in 1985 seventy-one years later, his recöllections proved astonishingly accurate.

From the 1880s on, an area of Japanese rooming and bath houses, shops, markets and restaurants grew up in Vancouver, centred around Powell Street near the Hastings Street sawmill. By 1900, this area was a thriving community conveying a strong flavour of Japan.

Today we would value such an area as part of the cultural diversity in our ethnic mosaic, but in the early part of the century the cultural differences generated suspicion and animosity. To the majority Caucasian population, the Japanese immigrants were mystifying — strikingly different in appearance, language and custom. Their diligence was undeniable, but their loyalty and commitment to Canada were a question mark. Restrictive convenants barred Japanese from owning property in many areas and a number of professions denied them entry. Japanese were effectively segregated by the chasm of culture and bigotry. They were themselves more at home in "Little Tokyo," as the Japanese ghetto was called; it was there they could fit in and feel comfortable.

My mother grew up on Powell Street and it was perhaps because of this that she had a much stronger sense of being Japanese than dad. She had gone to Japanese school and could read and write Japanese characters. She spoke a refined Japanese and had many

Japanese friends from those childhood days. Growing up in "Little Tokyo" had a strong effect on her.

Dad's childhood was very different. He grew up in Marpole, a small and predominantly Caucasian community. My father's family was one of fifteen Japanese families that had settled within a three-block radius. He went to an almost exclusively Caucasian school and although he must have stood out by virtue of his early inability to speak English, he was accepted and made lifelong friendships among the majority. English eventually became his main language. Though he was aware of the economic and social disparity between Japanese and white people, dad grew up always feeling he was Canadian.

When my father started to go to school, he absorbed the lessons being taught on Christianity. One day, dad told his father that if he was a good boy, he would go to heaven. "But," my father added, "you're a Buddhist and you will go to Paradise. Does that mean we won't see each other there?" This upset my grandfather so much that he started to attend an Anglican church to check it out. He eventually converted to Christianity.

MOTHER

My mother's entire life was circumscribed by struggle: the struggle to contribute to the family coffers while she was a teenager, the struggle to support her own young family during the Depression, the struggle to keep the family together during our incarceration in Slocan, and the struggle after the war to keep ahead of the bills. As long as I can remember, she was always there, hovering inconspicuously in the background when guests came over, and always cleaning up and putting things in place to keep the house from degenerating into chaos. She was the first to rise and the last to go to bed. Yet she was rarely an active participant in the conversations or discussions. She was just there.

My mother's life before I was born is a great mystery to me. She was born on April Fool's Day in 1910. Unlike my father who regaled us with stories of his youth, mom didn't talk about herself much. While my father has always unblushingly lived vicariously through me, I didn't learn until mom's death that she had been just as fiercely proud as dad of everything I did. She just didn't make a big fuss. She was a very typical Japanese female, allowing dad to be the centre of attention. She let him deal with all of the world's major problems while she went about the "less important" things like cooking, washing and running the household.

It was only after she began to exhibit the severe memory loss of Alzheimer's syndrome that mother's importance to the family became obvious to my sisters and me. Invisibly, she had always held our family together. She was forever the peacemaker. My father was volatile and often fired a devastating broadside — either verbal or physical — with the side of his foot on my rear. After such blow-ups, he would often come into my bedroom to apologize. I loved him all the more for it. It was only after mom's death that I found out that it had often been mom who had reasoned with him and had implored him to make up with me. Of course, all those years I had given *him* credit.

Mom was the baby of her family. Even now when I look at childhood pictures of her, I am awash with emotions at seeing her as a sensitive and innocent young girl. I only came to know her as an adult, after her life as a mother and wife were recorded in every stretch mark, scar and wrinkle on her body. She had been a shy and very serious girl. In high school, she had come in second at speed typing for all of Vancouver, and this remained the highlight of her public career.

Even though the Nakamura family lived in poverty on Powell Street in Vancouver's Japantown, mom remembered childhood as a happy time. After finishing high school, she went to work for Furuya, one of the oldest importers of Japanese goods to Canada. And it was there that she met dad. There was a strict rule against fraternizing between Furuya employees. The fact that the relationship was clandestine probably added a bit of zip to the romance. Dad eventually left so he could court mom openly. When they finally married, mom also left the company. Years later, the employers said that after mom left, they could never get the books to balance.

After the attack on Pearl Harbor, Japanese-Canadians were incarcerated in makeshift quarters in Hastings Park. I remember none of this. What little I do remember of the early part of the incarceration and evacuation seemed like adventure. My mother was just over thirty years of age. By April, 1942, dad had been sent off to a camp in the interior of British Columbia. All of our possessions and rights were suspended. With her husband gone, mom agreed to pack away all our possessions, close up the business and the house which were to be sold by the government, and leave Vancouver. We were on one of the first trains in June, 1942 to Slocan City, a ghost town from the days of the goldrush. Mom had to look after two six-year-olds — me and my twin sister — and a four-year-old daughter. Somehow, she managed to make it all seem exciting. The only time I can remember feeling worried was when our train was leaving Vancouver. Mother suddenly jumped up and dashed across the aisle, pushing a young child out of the way, to wave at people by the side of the tracks. It was completely out of character for her to be so pushy, but now I know why. Her parents were by the side of the track waving goodbye. When she sat down she was crying.

After the war was over, we were sent by train to Ontario and stayed in a hotel in Islington, just outside of Toronto. Dad set off to find a job, and after a month took an offer to work on a peach farm in Essex County. We joined him and my sisters and I attended a one-room school in Olinda while mom and dad worked. It was hard work for them at a subsistence wage; after a year, we settled in Leamington, Ontario, where dad and mom got better jobs in a dry-cleaning plant. Mom worked there part time.

My parents developed a great affection for the owner and his family who worked right alongside them. But the salary paid to my parents was barely enough to keep the family going. After four years, when he didn't get the raise he felt he deserved and it appeared the job wouldn't lead anywhere, dad told the owner he was leaving. The owner didn't bat an eye and simply said, "Okay." As soon as dad was gone, the owner zipped over to our place and told mom that dad was making a mistake and asked her to talk him out of it. But the owner didn't understand that she wouldn't do this. Mom never contradicted dad. Years later, the owner told my folks that his wife and daughter had urged him to put dad in charge of the plant and mom in the office. But he was persuaded by his card-playing cronies that he shouldn't trust a Jap. "Carr," he told my dad years later, "I made a bad mistake."

In Leamington, I started to work during the summers when I was eleven years old. My mother and I took jobs as farm labourers. She and I worked side-by-side with grown men, picking berries together. I could never keep up with her, nor could anyone else. These are my most vivid memories of her, her head shielded from the sun with a widebrimmed straw hat, picking berries, working on potato diggers and harvesting onions. We worked eleven-and-a-half hours a day, six days a week, and it was back-breaking work. I was always amazed to learn that many kids at school had mothers who stayed home all day. My mother came home from a full day of work, then did the household chores. She must have been very grateful that my sisters quickly assumed a lot of tasks around the house.

Dad's brothers had settled in London, Ontario where they had started a construction business. They did very well and dad finally decided to accept their offer to work for them as a cabinetmaker. It was a difficult decision for him, as the eldest in the family, to move to London and work as an employee for his brothers. But he wanted better educational opportunities for his children, and felt we would have them in a larger city. I continued to work in London, first as a labourer, then as a carpenter for my uncles.

Our financial affairs were always completely in mom's hands. Everything dad and I earned went straight to her. The same was true when my sisters started to work. My father could be incredibly impulsive and irresponsible; he would get something in his mind and decide that he just had to have it. Sometimes he would spend months of savings. Mom always bought or made my clothes, which suited me fine because I was never interested in fashions.

While I was familiar with dad's terrible temper, mom never hit me and seldom even raised her voice. She was very gentle and sensitive. Once, when I was a teenager, we were talking and I followed her into the basement where she was doing the laundry. Somehow she came to say that if anything went wrong when she and dad were older, I would have to help to take care of them. I started to tease her, using the sophistry of youth, saying I had never asked to be born and that she and dad had indulged themselves by having us. I went on to say that I had worked and done the best I could, but that once I was a man, I was on my own and I would have done my duty. I was feeling cocky and pleased with myself for being such a clever debater when, suddenly, I realized that my mother was weeping. My mother, who had never complained that life was rough, who had worked hard to keep the family going, had simply expressed the hope that she could depend on me. And I, all of fifteen years old, responded by tormenting her! I was sick with shame and never did that to her again.

When we moved to London in 1949, mom went to work as a bookkeeper and secretary for Suzuki Brothers Construction. She was the best they ever had: even-tempered, honest and efficient. I think she had her happiest years in London. She loved her home and had a sense of worth in the office.

Years later, when dad retired and they moved to Vancouver, she always talked of going back to her house in London if dad died before she did. She was lonely in Vancouver as dad rushed about fishing and going on trips. Mom had no close friends, although everywhere she had worked she evoked affection from people around. Her sister had been her great love outside her nuclear family. She adored Aki, whom she called "*Nessan*," meaning older sister. But Aunt Aki was in London and mom missed her terribly.

I worried about my mother and, in a moment of inspiration, I asked her whether she would like a part-time job as stockkeeper in my lab at the University of British Columbia. I kept hundreds of different strains of fruitflies which had to be carefully transferred and checked regularly. The pay for the job was low and turnover in personnel frustratingly frequent. The job was part time, but it gave mom a worthwhile routine. It was great for both of us. She was the best stockkeeper we ever had — reliable, careful and always cheerful. She would sit in a corner of the lab and was always included as one of the gang at parties. Often when students were getting boisterous and telling risqué jokes, I would glance over and see mom snickering away as she worked. Today, when I drop into the lab, I can still see her form hunched over trays of fly cultures.

After she passed sixty-five, the university began to put pressure on us to get her out of the job. We successfully fought each year to keep her, but as she approached seventy, she began to dread the battles to renew her job and she started to worry about her memory. I didn't realize she had Alzheimer's syndrome, but she sensed something was wrong and she was terrified that she might foul up one of the stocks. Finally, when her contract came up for renewal again, she said she had worked long enough. I didn't believe her, but we had to accept her decision. After that, I watched as she quickly slid into severe manifestations of memory loss characteristic of Alzheimer's. I'm convinced that her job held off the ravages of the disease, but we can never have a way of testing that.

When mom's memory began to go, my sisters, my wife Tara and I wracked our minds to find things to stimulate her, to make her life worthwhile and interesting. For the first time in her life, mom didn't have to worry about money; she could relax and do whatever she wanted. But what was there?

Mom lost interest in the jobs she had always done — sewing, cooking, housecleaning, the accounts. But her personality never changed. I always argued she didn't have Alzheimer's because she was so even-tempered, never nasty or mean. One positive change was that she did lose many of her inhibitions. She found it much easier to talk, and as so often happens in older people, she often remembered incidents from her childhood. She became openly affectionate with us. I loved to give her a big kiss and flirt with her. Her last years were the only times that I can remember her teasing dad in front of us and she loved it when I told slightly off-colour jokes.

Dad became the focus of her daily life and increasingly mom consumed all of his time. Towards the end, she would occasionally look up at dad and ask when he was going home. It must have frozen his blood. When he asked why, she might reply, "My husband's coming home soon." When dad would ask who she thought he was, mom would then guess, "I think you're Frank" (her brother) or someone else. Mercifully, lapses like that didn't happen often or last long.

I lived in dread of my mother becoming incontinent and losing interest in her personal hygiene. We feared the day when she would not recognize anyone. Without memory, we lose the connections that are so vital for maintaining a place as social animals. It's really the past that defines us; without it we only *appear* normal, while in fact we have lost a big part of our humanness. Mom's death from a massive heart attack was a shock because she had seemed so strong and healthy and it happened so suddenly. But along with the grief, I cannot deny that I felt relief. The fact that she had been resuscitated and placed on a respirator is what made her death grotesque.

For seven days after her heart attack, mom's body continued to function without the aid of any machines. She was removed from the intensive care unit a day after she had been resuscitated and was placed in an ordinary hospital room. My father, sisters and I kept a vigil in that room twenty-four hours a day until her "death." We slept on mattresses nurses put on the floor for us and ate in the hospital cafeteria.

I spent many hours just sitting there looking at the form that was no longer my mother. This valiant figure had laboured in fields, struggling to feed and clothe and protect us. To the end of her life, mom was the source of love and warmth. And since she died, life has never been the same for me.

FATHER

As far back as I can remember, my father always seemed like a titan to me. With his enormous exuberance for life, he was garrulous, generous and inquisitive about everything. His vast faith in me gave me the nerve to try what my own inclinations often resisted.

My father was my inspiration, my hero, my model. And just as he was also cantankerous, opinionated and narrow-minded, so too, I have inherited those qualities — and continue to struggle to exorcise them. But he taught me some of the most worthwhile lessons in life. He showed me how to snell a fish hook, where to cast a fly in a stream, how to build a lean-to in a blizzard, how to find hellgrammites for bait, and the way to right a canoe when it tips. When I look at a river and say, "That's beautiful," it means that it looks like a great place to fish. That's my dad's influence. He taught me to reel with my left hand so I wouldn't have to change hands to cast all the time. Ever since, anything having to do with winding, like using a can opener, I do with my left hand. He taught me that boys don't swear in front of women. He taught me the aphorisms that I continue to use and repeat to my children:

"Whatever you do, do it with everything you've got — whether it's scrubbing the floor, fishing for trout or studying for exams."

"You are what you *do,* not what you *say.*"

"If you're going to stand up for anything important, you will make enemies too."

My father was born in Vancouver in 1909, the first son in a traditional Japanese family. In spite of being a native-born Canadian, dad spoke no English when he started school. His given name was Kaoru, but when he went to register, the teacher couldn't spell his name and wrote down Carr. The friend who had taken dad to school said "That's all right," and Carr has remained his name ever since.

He was thrown into grade one and, not knowing any English, had to learn by imitation. Once he turned around and the boy behind him dabbed paint on his nose. Dad thought that was great fun and dabbed the boy back. In no time at all, they were both covered in paint. The teacher hauled them both into the cloakroom and strapped them. It didn't hurt but the other boy started to wail. So dad did the same. When they were left alone, the little boy would sob and then every so often stop and peer out the door to check the action. Dad copied everything the boy did.

At school, dad didn't know he was supposed to ask permission to go to the bathroom and so he just got up to go on his own. One day, the teacher hauled him back into the room because she thought he was being mischievous. She punished him by making him stand up beside his desk. Finally, dad couldn't hold back any longer and wet himself. He watched in horror and shame as this stream of urine flowed all the way to the front of the room. After that, dad was allowed to leave on his own whenever he wanted.

Total immersion worked for dad and English very soon became his main and preferred language. Unlike many *Nisei,* he became very articulate and outspoken in the English language. He has remained fully bilingual all his life.

Dad's parents were totally preoccupied with eking out a living and providing for the material needs of the family. They had no time to teach him about the world around them. That was for dad to pick up as best he could. He was fascinated by nature from early childhood and he remembers exploring his neighbourhood when he was about four, and finding a creek full of stickleback fish. He found he could dip them out with a net and so he kept putting them into a glass jar until he had literally filled it with fish. When he got home and poured them out, he was astounded to find they were all dead. That was when he learned that fish need lots of water to live.

One time, before he started going to school, dad went out on the salmon boat with his father. They docked on an island and while his father carried out his business, dad explored around the dock and discovered arbutus trees. These are the hauntingly beautiful trees which grow on rocky ledges on the West Coast. Their distinctively rust-coloured, smooth-barked limbs are usually twisted and gnarled by the wind. Even at that age, dad was struck by the beauty of the trees, so he dug up a number of small plants and put them in a bucket to take home. When his dad arrived and found the bucket, he simply dumped the plants overboard and got on with work! His dad had no time for esthetics. But my dad's love of trees — and all plants — has lasted through his life and the breadth and depth of his knowledge have always astounded me.

Dad must have spent a lot of time on his own during his preschool days. He has vivid memories from those times. When I was doing an honours degree in biology at college, he mentioned that as a child he had watched bumblebees flying to their nest. "You know," he told me, "while I was watching them, I was surprised to see that there were tiny creatures on the bumblebees' bodies that looked like scorpions!" It had made a lasting impression on dad. Well, I thought I was a hotshot budding biologist and had never heard of such things. So I think I nodded patronizingly in order to humour him. I was amazed to discover, while in graduate school, that there are indeed parasites of insects that are called "pseudoscorpions"! I've had the greatest of respect for his observational powers ever since and constantly think of that story to remind myself about the dangers of being a know-it-all.

From a very early age, dad had to do chores around the family farm. It meant he never had time to play sports and other extracurricular things at school. When he was about eleven years old, he went to work for a Caucasian family as a houseboy. He lived with the family, receiving room and board and five dollars a month (he eventually saved enough to buy a bicycle). In return, he helped the housewife in the kitchen (and learned to be an excellent cook), did the house cleaning, went shopping and kept the yard tidy. He doesn't remember it as an oppressive job and was treated very well.

During his spare time, he took advantage of his employer's

set of *The Book of Knowledge* and read the entire encyclopedia from cover to cover. Years later, before I started school, I loved to stand behind dad's pressing machine as he worked and talk to him. I would assault him with all kinds of questions and he used what he had gained from *The Book of Knowledge* to satisfy me. He never made anything up and always told me when he didn't know something. But that wasn't very often.

Dad worked at a number of jobs as he grew up. He helped his father fish for salmon. He was a fish buyer for the canneries. He sold clothes for a Japanese clothing company throughout Alberta and British Columbia. He travelled a lot and would get off the train at Kamloops and visit men who were down and out, and lived in "hobo jungles." He was struck by the fact that Chinese farmers who grew vegetables such as tomatoes would always plant a few rows of potatoes for itinerants to help themselves. I guess it was a self-protective act, but nevertheless, a generous one. In the actual hobo camps, there was a kind of code men went by so you would always find a can to cook soup in and dry wood left by the people before. In Vancouver, dad would watch the freight trains come in and see dozens of men riding the rails and jumping off before the locomotives arrived in the freightyard. Ever since I was a child, dad would tell us about his experiences with hobos; he made it sound like a romantic way to live.

When dad met mom, one of the first things he did was talk to mom's folks to ask permission to date. Since mom's dad had descended from the *samurai* class, he was a pretty imposing man. I can imagine dad shaking in his boots as he made his pitch. But dad's Japanese had deteriorated, and his being nervous didn't help. So he was amazed to hear mom's dad answering his request for dating privileges by telling him, "You're too young to get *married.* If you're serious, come back in five years and I'll consider it." Later, dad dazedly asked mom whether he had actually said he wanted to get married and she confirmed that this was how it came out. Dad hung in there and five years later went back to get permission to marry.

Mom's older sister had been married by arrangement. Dad and mom represented the new generation. They had fallen in love and courted like Canadians. Theirs was a very physical relationship and other Japanese Canadians to whom I've talked have been astounded that my parents were openly demonstrative. As early as I can remember, my parents would actually "neck" and they kissed whenever dad left for work and came home. They had a very passionate relationship that lasted right up to the time of mom's death.

After they were married in 1934, a dry-cleaning shop owned by a Japanese man came up for sale in Marpole, right near my grandfather's place. Grandfather wanted to keep his family close by, and urged my father to buy it. He offered to put up the money and so dad went into the dry-cleaning business.

Dad was a very hard worker all his life. But he played as hard as he worked. He always had a great love of the outdoors; his brothers built a pleasure boat for him and he loved to take mom on trips on the ocean. He was an avid fisherman and very good at it. He had a group of fishing buddies whom he would go off with on weekends, while mom would stay home and take care of the kids. Shortly before the war started, one of his good Canadian friends advised him to stop fishing. "Carr," he said, "when you're up there fishing in the bushes, someone is going to come along and see a Jap. If he shoots you, you'll be dead and he'll be a hero!" It was good advice, but it broke dad's heart.

For years I wanted to take dad and mom to visit Japan. Other than dad's month-long stay there when he was five years old, neither had been to Japan. But dad was adamantly against going, and didn't go until after mom died. I think part of the reason for his hesitation was that the Japanese he spoke reflected that his parents were uneducated and came from a very low class. Furthermore, people in Japan are very quick to adopt new words and today's language is full of English cognates. The Japanese that dad speaks is a century old and low class. Dad always felt that Japanese visitors he met looked down on him for the coarseness of his language. When dad finally did go to Japan, he found that while Japanese people realized immediately that he spoke a different kind of Japanese, they were intrigued and curious. He became a centre of attention rather than an object of contempt.

As I watched the relentless deterioration in my mother's memory before her death, her greatest gift to me was a new version of my *father*. Through my worshipping child's eyes, he had always been a colossus among men. As adults, my sisters and I had become aware of how crucial mom's role in the family had been and we were much more critical of dad's domineering ways. So his lustre and stature were inevitably tarnished. But as my mother's interest in cooking, bookkeeping and housecleaning faded, my father, who had never bothered with such things, took it all on. And more than that, he became inseparable from my mother, her companion, nurse and protector. "She worked hard all her life for me," he would say, "and now it's my turn to make it up to her."

He did it without any fanfare and in the process revealed a new dimension of himself. He took mother everywhere. He bought her fashionable clothes, scrubbed her back and feet and talked to her constantly, even though she would promptly forget what he said. He enjoyed her company and they made a touching sight, arm in arm as they walked down the street with their dog. They walked that dog together every morning and evening, rain or shine.

My dad fiercely resisted my efforts to hire someone to give him some time to himself. When he finally relented, he was astounded at what an all-consuming occupation mother had become. I saw depths of selflessness, compassion and yes, even at his age, love and passion I never thought he had. And on occasions when I dropped in to find him weeping with frustration, anger and grief over mom's condition, his human dimensions grew immensely. In the end, my mother left me with a priceless gift — a father who has, in his humanity, far surpassed his mythic proportions of my childhood.

As long as I've known my dad, he has been an agnostic, without a belief in God or in an afterlife. In recent years, he has looked back at his Japanese cultural roots and claims to be a Shintoist — a worshipper of Nature. When mom's memory began to fail and dad got cancer, he thought that he might die before she did and worried about what would happen to her. So he talked at length about joint suicide with mom.

People in the West find the Japanese acceptance of suicide repugnant. I think it's because we (I say we because I am really a part of the West) have put such a high value on individual life. We believe that any child can become prime minister or president, and that the sun rises and sets on *us*. So when we die, the universe winks out.

The Japanese have a very different perspective. The "Japanese People" are considered to be a single entity, having a greater importance than any individual. Just as individual cells make up a whole organism, so all of the people together make up Japan. The Japanese see their country in an historical perspective, as a river flowing through history. At any given moment, individuals are a part of the current. So *kamikaze* pilots or *harakiri,* which seem so extraordinary to us, are perfectly acceptable to the Japanese way because individual existence is never as important as the continued survival of Japan.

The first time I really believed that my father didn't fear death the way I do was when my grandfather died. I knew how much dad had always wanted and needed his father's approval. Because I can't imagine life without *my* father, I thought dad's world would crumble when his father died. But it didn't. Dad has long operated on the belief that it's what you do with people *before* they die that matters. So he will go to extraordinary lengths to do favours or go to see people who are very ill. Once they die, he doesn't feel regret or guilt for things he *should* have done. Thus, when mom died, his relief that it had happened quickly and before she had lost her dignity was far stronger than his grief. His own mortality and imminent death do not terrify him. It's that acceptance of death that has let him take chances throughout his life.

I, on the other hand, have feared death as long as I can remember. As a life-long atheist, I have dreaded, not the process of dying, but the terrible consequence of not *being* forever after. But the accumulation of experience with age — my visible physical aging, the deaths of my mother, friends and acquaintances, the maturation of my children, reaching fifty — has dulled the pangs of my own mortality. I now feel a visceral resonance with the Japanese attitude toward death that renders me more immune to criticism for the things I do. Death isn't as frightening because I take myself less seriously.

Since we must die like all other life forms, I take great comfort in the idea that life goes on, that my children and nature persist after our deaths. As psychiatrist Jay Lifton has pointed out, that is what makes the massive species extinction going on today and the prospect of nuclear war so horrifying. Human activity now puts not only our own species in jeopardy, but much of nature itself. A new and terrifying dimension has been added to the prospect of death.

Though dad has always been my hero, during my adolescence I figured I knew everything I needed to go out on my own. I found dad's standards and demands too exacting. I resisted him, and like so many teenagers I chafed under his rule. But then remarkably, when I was in my late teens, dad came to me and said that I was old enough to be a man. From then on, he told me, I should make my own decisions and he would not interfere. In hindsight, I see that as a great gift to me because he stuck by his word. He was always there when I wanted advice, which was often. But he never pressured me or tried to get me to do what he wanted. When I decided to get married, I knew he thought that I was too young and inexperienced. But he didn't object and supported me fully when he saw I was determined.

As I progressed through my stages of growing up, dad and mom exulted with the completion of each step. I was the first of the Suzukis to complete university, to go on for a graduate degree, to get a Ph.D., to become a university professor. I became dad's surrogate ego. My achievements became his — a validation that my father had achieved something. I have never felt that dad needed that kind of certification because he is an incredible man on his own. But I've never begrudged his enjoyment at being my father.

Well, I am occasionally embarrassed because he is so gregarious and quick to strike up conversations with anyone within earshot. It seems to take one nanosecond before he has somehow managed to ask a person whether he or she has ever heard of David Suzuki and to announce that he's my father. Meanwhile, I cringe in a corner and pretend I'm not with him. I meet all kinds of people who tell me they've met my father somewhere and have heard all about me. He's like my roving ambassador who basks in reflected glory.

Dad is also my greatest critic. He consumes everything I do. He listens to as many of my radio shows as possible, reads everything I write and watches all of the television programs. He seldom raves about anything I do, but he often criticizes me when something didn't make sense, or was boring or too technical. I try to think of him as my audience whenever I write something. He also has a broad vocabulary and quite a few times has pointed out that I've misused words. I value his ability to be constructively critical even while he also remains my biggest fan.

As dad and I grow older, I come to appreciate him more and more. Whenever I did something dumb as a teenager, he would call me an "educated fool" and the words stung because I knew he was right. I was good in school, but what use was that if I couldn't do the practical things which are important in life? I don't know how to use dope to waterproof a fly line, sharpen a saw, or fix a car motor. I can't mend a leaking pipe, steer a boat through ten-foot waves or carry on a conversation in Japanese. Dad can.

It is unfortunate that society has come to ascribe such economic and social status to higher education. One should feel pleased at having acquired a good education, but should never equate that with superiority over those who haven't. I could never match all that my father knows, even though he has little formal education.

It is sad that today as technology affects our lives in ever greater degrees, *human* obsolescence has become one of its costs. Stored in dad's brain is a wealth of observation, experience and inventiveness that relates to a different time when people lived closer to the land. In today's high-tech world, dad's fund of information is seen as nickel knowledge. But when young people meet dad and spend time listening to what he knows, they are invariably drawn to him and filled with excitement. We need the wisdom and experience of our elders.

The women on Dad's side — his mother (centre), paternal grandmother (left), maternal grandmother (right).

Dad's father mending the nets for salmon on the Fraser River.

(right) The Nakamura family. That's Mom, the baby in the front centre with Uncle Frank on the left and Aunt Aki on the right. Between my grandparents is Uncle Sally.

Dad had all the good looks of his father.

Mom (sitting) and Aunt Aki made a smashing pair.

Uncle Sally before he left for Japan.

Mom and Dad's wedding on March 21, 1934.

CHAPTER TWO

A New Generation — Childhood to War's End

IN JAPANESE SOCIETY, the birth of twins is not a happy event — it is frowned upon. The first child to appear is considered the younger of the pair. The Japanese believe that out of politeness the elder steps aside, so to speak, to let the younger out first. So my twin sister, Marcia, having been delivered minutes after me on March 24, 1936, is considered older than I.

Irrespective of my chronological status, as a male I occupied the top rung in the family pecking order as the family filled out with two more daughters. To my father, in typical Japanese fashion, I was the favoured and most important child, the only son. It was a burden my three sisters had to bear all their lives. While dad would go fishing and camping with his male chums or me, it would never occur to him to take my mother or sisters.

What is remarkable is that my sisters accepted this hierarchy. They never questioned my right to be so highly regarded, though I had done nothing to deserve it. I never heard them complain that I was getting special favours or ask why I didn't pull my weight. They had to fight to express themselves as individuals and in the process each became a special and tough person in her own way. But those early years of social conditioning still persist. My wife marvels at how my sisters, now mature women, continue to rush around and look after my creature comforts whenever we drop in for a visit. And I too very easily slip into the comfortable familiarity of the old role. That's why, while I struggle to live up to being a feminist, I would never claim to be "liberated." I've lived most of my life with the cultural assumptions of male dominance and they are not exorcised easily.

By the 1930s, over 75 percent of the Japanese in Canada were living within a seventy-five-mile radius of Vancouver, mostly clustered around Powell Street in Vancouver, the fishing village of Steveston, and other areas around lumber mills, canneries and farms. Our family, however, lived in predominantly white Marpole. Being Canadian or Japanese was not a matter I thought about in the five years we lived there; our family was assimilated into that village as Canadians. English was the only language spoken in our home.

We lived in the back of our dry-cleaning shop, next door to our Canadian neighbours, the McGregors. Their youngest son, Ian, was my age and my best friend. One day a new boy who had moved in down the street approached Ian and me. The boy told Ian not to play with me because I was a Jap. At that point I shot back, "But I'm a Canadian, just like you. I speak English, don't I?" He reluctantly agreed. "I eat the same kind of food as you." Hesitant acknowledgement. "My clothes are the same as yours." "Well, I guess you're right," he finally admitted. "You must be a Canadian — but you still *look* like a Jap to me."

There were some traditional Japanese obligations which we observed — family get-togethers, meals and holidays. I remember our family celebrations on Boys' Day in May, a Japanese festival where miniature warrior dolls and costumes and swords would be laid out on a tiered altar. Grandfather had bought a very elaborate set of Japanese dolls for me and another for Marcia to be used for Girls' Day in March. These had been gifts at our birth and dad was relieved that his father had accepted the birth of twins so well. I loved these ornaments and their related festivities because they had such an aura of specialness. I never knew it was a Japanese celebration, any more than English children think Guy Fawkes' Day is an English event, or Québécois kids know St-Jean Baptiste Day is French. They just have fun because that's what festivals are for.

Because I was just over six when we moved away, I only have a few memories of Marpole. In my childhood, ice boxes and pantries were our refrigeration while horses were the engines for most wagons and carts. When the ice wagon made deliveries, we would run out on the street to pick up pieces of ice that dropped off. It was a great treat to suck on ice in summer. I remember being amazed when dad dug into the sawdust bin he had in the back of our house and hauled out a block of ice that he had stuck there in winter. I didn't understand the concept of insulation, but sawdust's ability to keep ice from melting made a lasting impression. As an adult, I've carried frozen fish back and forth across Canada. From dad, I learned to wrap them in many layers of newspaper — a reminder of the insulating properties of wood.

Almost all of my early memories involve fishing and camping with my dad. My first fishing trip was to a small lake called Loon Lake, located in an area that is now a forest endowed to the University of British Columbia. Loon Lake was and still is full of trout stunted in size by their density. I found that by simply dropping a worm straight down under the boat, I could easily catch them. My first time out I caught my limit of fifteen and have been an avid fisherman ever since. There were so many fish in that lake, when a bare hook was cast out, fish would hit it as it was being retrieved.

As a parent I now look back on that early fishing trip as an important lesson. It is crucial that a first outing be fun and that the objectives are met whether fishing, camping or canoeing. Nothing can turn a child off the outdoors as quickly as having a terrible experience the initial time out. I'm always sad when I see a child with a cheap dime store fishing outfit, huge hooks and no chance of catching fish, sitting disconsolately by a lake or creek. It's no fun and it sets one's entire attitude towards fishing, often for life.

My father always loved to take boys fishing and camping. He believed that such trips were experiences that put youngsters in touch with nature and kept them out of trouble. He would often bring along kids who had had scrapes with the law or were having trouble in school or at home. Even today, I am approached on streets and in airports by grown men who tell me dad took them fishing when they were young and what a memorable experience it was.

In Marpole, dad frequently took me with him to the sporting goods shop to check out the fishing gear and exchange fishing gossip. One day he looked up to see all of the clerks crowded around me as I reeled off the names of patterns of flyhooks. I had learned them by watching dad tie them. He was tickled, but hustled me off to keep me from getting conceited.

Dad also taught me how to clean fish at a very early age, and I still love the job. Cleaning fish is a direct physical contact with nature that I value. I've never understood why it is considered such an onerous task.

We bought our first tent in Vancouver when I was four. It was specially made from sail silk with a bottom sewn on it. It was just a little pup tent, but I loved it. We bought a down sleeping bag that was on display in a store and dad and I crept into it and lay out full length to test it on the spot. On weekends, we would drive to a favourite spot and park the car. Dad would put on a pack, then load me on top of his shoulders and off we would go.

One night after work, dad drove to Loon Lake and we hiked up in the dark. I held the flashlight and remember dad complaining because, like most youngsters, I kept aiming it all over the place while he stumbled about trying to see the trail. As we were climbing a hill, Sport, our dog, started to growl. We turned the flashlight around and spotted a glistening pair of eyes. It was a bear silently watching us, and in the glare of the light it slowly turned and went away. Throughout my childhood, dad taught me to love and respect nature, but never to *fear* it. Seeing the bear was a fascinating and exciting experience.

Some weekends we would drive out to Chilliwack in the Fraser Valley to fish in the Vedder River, which is still famous for its steelhead trout. At the end of the road, dad would hire a horse from a man he knew and we'd ride it several miles upstream. Then we would let the horse go to make its way back on its own. We would drop our gear, fish a while, then pitch the tent and camp overnight. The next day, we'd fish our way down the river. I don't remember owning a rod back then, but I loved to watch dad catch fish while I played on the river bank. Even now I get as much pleasure watching others land fish as I do catching them myself.

On one trip up the Vedder, it was so hot that dad told me to cool off by jumping into the river. "You mean with my clothes and shoes on?" I asked in disbelief. "Sure," he replied. So I did and had a marvellous time. It's wonderful to get wet or muddy with clothes on. It has always amused me to see each of my children respond when I've let them muck about in water with their shoes and clothes on, the same way I did when I was their age.

Those early experiences camping and fishing shaped my interests for the rest of my life. I started camping with all of my own children when they were a few months old. It's very different now — we usually drive to a well-groomed campsite with a fireplace and wood — but it's still being out-of-doors. I firmly believe this is a vital experience for youngsters. It reinforces a connection with the environment and nature that cannot be experienced in urban settings.

RACISM, WAR AND MASS EVACUATION

While I was living the innocent life of a child in Marpole, the drums of war were beating loudly in Asia long before the surprise attack at Pearl Harbor. Everyone could feel it coming and the increase in inflammatory rhetoric against the Japanese-Canadian community could be seen in the newspapers of the early 1940s. Long before Pearl Harbor, fear and hatred of Japanese had been rampant.

In the last two decades of the nineteenth century, Japanese immigrants flowed steadily into Canada. By 1901, there were

4,738 Japanese registered in the country; 97 percent of them concentrated in British Columbia. Japanese and the more numerous Chinese represented 10 percent of British Columbia's total population. By the turn of the century, almost 2,000 licensed commercial fishermen (out of 4,722) and a quarter of the labourers in the lumber industry were Japanese. While Orientals were valued for their diligence and willingness to work for much lower wages, they evoked the classic response among the groups with whom they competed. They were thought to have far fewer physical needs and were therefore unfair competitors since they worked for less money. Separated from the Caucasian majority by race, culture and language, Asians were readily perceived as a sinister threat to British tradition.

Immigration from Japan accelerated and reached a peak in 1907. Rumours flew that fifty thousand Japanese were being brought in by the Canadian Pacific Railway. On September 7, 1907, a mass parade was called to protest the influx of Asians. It grew to a crowd of eight thousand who were whipped into a frenzy by a succession of speakers. The crowd then flowed uncontrollably from City Hall into Chinatown on Pender Street. The Chinese population retreated behind locked doors, but as the mob surged toward Little Tokyo on Powell Street, the Japanese chose to fight. A riot broke out in which considerable property damage and a number of individual beatings occurred, but no lives were lost as riot squads imposed an uneasy truce.

The future prime minister, MacKenzie King, was called upon to head a Royal Commission to investigate the riot. He settled property claims and negotiated a voluntary agreement from Japan to limit emigration to four hundred a year, thus allaying fears of a massive buildup of Japanese in British Columbia. Ironically, a large proportion of the subsequent immigrants were "picture-brides" — arranged marriages — who added social stability to the Japanese community, but served to increase Caucasian anxiety as babies were born.

The British Columbia legislature repeatedly tried to restrict Japanese entry into the fabric of society, only to be foiled by the federal government which was anxious not to jeopardize Anglo-Japanese relations. The provincial government attempted to raise the residency requirement for naturalization of Orientals from three years to ten, but the federal government overruled it. British Columbia acted to prevent Orientals from working on provincially sponsored projects, but again, Ottawa overturned it. British Columbia tried to restrict the number of immigrants, but bowed to federal pressure. In 1922, British Columbia unsuccessfully appealed to the federal government to amend the *British North America Act* to prohibit Asiatics from acquiring proprietary interests in farming, lumber, mining and other industries.

The province did extract a promise from Ottawa to eliminate Orientals from fishing by reducing the number of licenses issued each year. As one of the leaders in the fishing community, grandfather Suzuki was elected to make the decisions as to who would keep — or lose — their licenses. It was a thankless task and each year he would receive threats on his life. By 1925, the Japanese had lost almost half of their licenses. Ironically, as fishermen were forced to seek other means of employment, they diversified and became threats to white Canadians in lumber, farming and business.

The diligence of Asians was seen as a threat to take over Canada's economy. Many immigrants had been farmers in Japan and they found vast tracts of uncleared land in British Columbia. They set to work clearing land by hand. Applying efficient methods learned in Japan, Japanese soon dominated the berry market. As might be expected, the tightly knit, hardworking Japanese families were mistrusted by their farming neighbours.

For the Japanese, the greatest impediment to economic mobility was the denial of the right to vote. Entry into the ranks of many professions and businesses required that a person be listed on the voters roster. But no Asians were allowed to vote and by the absence of all Japanese — naturalized or Canadian-born — from these lists, they were ineligible for jobs in hand-logging, law, pharmacy, the civil service, forestry, the post office and public health nursing, among others.

In the attempt to gain the right to vote, *Nisei* were caught in a catch-22 position. They wanted full and equal rights of citizenship, but since there was no comparable pressure from the much larger Chinese community, the Japanese were seen to be aggressively pursuing ulterior motives. Provincial Member of Parliament Thomas Reid saw Japanese-Canadians acting as mouthpieces for the Japanese government. Granting them the franchise in British Columbia, he believed, would be the foot-in-the-door for the Japanese government who would then have "an active voice in Canada and so help to shape the policies of this country." How this could be achieved by a minuscule portion of the electorate was never specified.

Rampant racism was often a consequence of "biological determinism," a belief then popular among geneticists that human behaviour is dictated by heredity. Thus, B.C. Member of Parliament A.W. Neill stated in the House of Commons on March 16, 1937: "To cross an individual of a white race with an individual of a yellow race is to produce, in nine cases out of ten, a mongrel wastrel with the worst qualities of both races."

This was a crude expression of what geneticists had been claiming for decades — that offspring of interracial marriages result in "disharmonious combinations." The rationale was that since different races had evolved in isolation and were adapted to very different environmental conditions, intermixing their genes could only reduce these adaptive qualities in *both* environments. We now know, of course, this is nonsense.

Neill clung to his hereditarian beliefs, as an exchange with Prime Minister MacKenzie King in February 1941 shows: "Whatever the opinion of the Prime Minister or of the government here, we in British Columbia are firmly convinced that *once a Jap always a Jap.*" (my italics)

Few would state as baldly what has been a biological justification in numerous social debates from Natives to blacks, Québécois, the poor and welfare recipients.

From early in the twentieth century, a new generation of Japanese-Canadians like my parents had rapidly grown in number. These *Nisei* were radically different from their parents, linguistically, culturally and socially. The original immigrants, the *Issei,* saw in them a corrosion of the traditional Japanese values of filial piety, modesty and hard work. Even though the *Nisei* were outstanding students, their parents disapproved of dancing, dating and their perceived "laziness." When Japan invaded Manchuria in 1931 and proceeded to bloody China's nose, the *Issei* were proud. They held drives to send financial and material support to Japan. This riled the Chinese community and their mutual animosity deflected the two groups' energies away from their common foe — the white bigots of British Columbia. Just as Hindus and Sikhs in Canada continue to fuel the rivalries generated on the Indian subcontinent, so too the Chinese and Japanese remained enemies on the West Coast.

When war was declared in Europe in 1939, *Nisei* tried to volunteer for the Canadian Armed Services, in large part to demonstrate their loyalty. But the federal government, under heavy pressure from its B.C. constituents, denied *Nisei* the right. Ironically, *Nisei* were willing to lay down their lives to protect democratic ideals that were being denied them.

B.C.'s *Nisei* knew they were in for a rough time. But in spite of the racist rhetoric of politicians and the media, they clung to their faith in Canada and Canadian fair play. At worst, they expected they might have to obey curfews and live under greater surveillance, or perhaps to report regularly to military offices and suffer the kinds of restrictions that had been imposed on Italian and German nationals in Canada in 1939.

The *Nisei* hadn't anticipated the treachery of the devastating "sneak attack" on Pearl Harbor on December 7, 1941. That attack confirmed every bigot's belief in the inherent deceit and untrustworthiness of the Japanese race.

On the very first day of the Pacific war, thirty-eight "hard-core" Japanese nationals were arrested and subsequently interned. Crew members of fishing boats had long been rumoured to be charting the B.C. coast for the Japanese navy. Some boats were claimed to be disguised warships and part of a spy network. Steps were immediately taken to identify and register the 1,200 boats belonging to naturalized and Canadian-born Japanese (in other words, Canadian *citizens*). They were eventually impounded, and although illegally seized and held the boats were soon sold by the authorities at bargain prices in order to get the vessels fishing again. Fifty-nine Japanese language schools and three Japanese newspapers were shut down on December 7. All Japanese nationals (i.e., aliens) were required to register by February 7, 1942 and to sign a declaration of good conduct and indication of domicile.

Ironically, the place Japan chose to attack — Hawaii — was also the home of 160,000 people of Japanese ancestry, fully 38 percent of the islands' population. If there was ever a place where an extensive Japanese fifth column could flourish, Hawaii was it. Yet no verified case of Japanese-American treason in Hawaii was ever reported and no evacuation, incarceration or even curfew was imposed! In contrast, 113,000 Japanese spread out along the coasts of California, Oregon and Washington were evacuated.

But in matters of racism, Canada didn't have to look for American examples. Two weeks before the United States began to remove Japanese from the West Coast, British Columbia had already begun. In fact, Americans pointed to the British Columbia example as a model for *their* operation.

The shock of Pearl Harbor and the enormity of the damage inflicted on the U.S. Navy imposed a strange calm for a week. The Japanese community in British Columbia knew something would come, but few anticipated the intensity of public fear, hostility and opportunism. When the storm finally hit, it would devastate the community, pit generation against generation, create deep rifts between groups, permanently disrupt the tightly knit ghettos and inflict psychic scars that have yet to be fully diagnosed.

The hurricane struck on December 16, 1941. *All* Japanese-Canadians, regardless of birthplace or naturalized state were required to register with authorities. From that point on, the distinction between Japanese nationals, naturalized *Issei* and Canadian-born *Nisei* would disappear. Bigoted comments like those from Neill had prevailed and the government acted as if "once a Jap, always a Jap." Along with mandatory registration, any Japanese person could be asked to show a card at any public assembly. Japanese were marked.

It was an anxious holiday season for the Japanese spotted along British Columbia's lower mainland. On January 8 and 9, 1942, a meeting of politicians, the military and RCMP took place in Ottawa to assess the Japanese "threat." The records clearly show that both the Armed Services and the RCMP were confident that quick action had nullified any potential threat from the Japanese community. But in British Columbia, people felt exposed and vulnerable to Japanese forces. The Japanese Imperial Army seemed to be invincible as it piled one military success upon another in those early months.

On January 14, evacuation of a larger and more vaguely defined danger was announced. All Japanese males between eighteen and forty-eight years of age were to be removed from the West Coast by April 1. Thus, the aged, the women and the children would be left behind, a prospect that created enormous anxiety in the Japanese community.

Along the United States coast, most of the 113,000 Japanese were American citizens. On January 29, 1942, the U.S. Department of Justice began to move them away from the West Coast. A day later, Canada defined a protected zone from which all "enemy aliens" were to be excluded. It was a strip from the Pacific to about one hundred miles inland and extending from the U.S. border to the Yukon.

On February 19, the United States formally announced its plans to evacuate all Japanese from the West Coast. Meanwhile in Canada, it had taken a while to get the Japanese nationals ready to leave. On February 23, the first group of one hundred men, the vanguard of 1,700 who would eventually be uprooted, left Vancouver. On February 24, Minister of Justice Louis St-Laurent was empowered to control the movement of Japanese. Two days later, plans for a mass evacuation were announced. That day, it was indicated that a curfew was imposed and cars, radios and cameras belonging to Japanese would be confiscated.

A week later on March 4, the B.C. Security Commission was set up to plan, supervise and direct the evacuation. In addition to these responsibilities, the Security Commission acted to police Japanese conduct, activities and discipline. The Commission was in charge of housing, food, health care and protection. Within ten days, the Security Commission made its first move, bringing people into detention at the livestock buildings of Hastings Park. Housed in temporary quarters set up in cattle stalls and horse stables, the people were herded like the animals they displaced.

For the Japanese community, the war brought into sharp focus all the divided loyalties of a bicultural group. It was the final test of their allegiance to Canada. To the *Issei,* the war pitted the land of their birth and culture against a country that had not welcomed them in the first place. Undoubtedly, there were those whose loyalties were with Japan, but they were a small minority. The overwhelming response of the *Issei* and *Nisei* was a hardening commitment, however painful, to Canada.

Suspension of civil liberties, confiscation of property and incarceration were made possible by the *War Measures Act,* a deplorable piece of legislation for a democracy. It confers upon the Cabinet of the federal government sweeping powers in the event of war, invasion or insurrection. In the interests of the security of the country, these powers allow censorship, control over publications and property, and the arrest, detention, exclusion and deportation of anyone — all without any burden of prior evidence or proof of guilt. It is one of the dilemmas of a democracy that the ideals of equality, and freedom of expression and movement only matter in times of stress. Yet the *War Measures Act* renders those rights void at the very time they are most precious. It had been invoked before and would be again. But the application of the *War Measures Act* against the Japanese was the only time it was used on such a scale against an entire group. It would never again be kept in force for so long, and it was applied without a shred of evidence to justify it.

In Canada, all able-bodied Japanese men were separated from families and sent to road camps, while their families were evacuated to remote areas in British Columbia's mountains, to "ghost towns" which had once supported thousands of people searching for gold and silver at the turn of the century. Slocan City, Greenwood, Kaslo, New Denver, Sandon — these were to become home for over fifteen thousand people. The settings were dazzling — the Slocan Valley lies between spectacular mountains of the Selkirk Range. The mountains bordering Slocan Lake on the west have recently been designated as Valhalla Provincial Park.

An advance contingent of men was sent to construct housing for the flood of evacuees to follow. By early June, the first families began to trickle into the camps. The first arrivals were housed in old buildings that were still standing, and in tents. Later, three-room houses were thrown up to hold two families each. By 1943, more than twelve thousand people were living in these camps, over a quarter of them in the Slocan Valley, seventy miles north of Trail. There had been little advance planning, and the Security Commission struggled with the enormous logistics of such a large move.

Nothing reflects the lack of foresight and planning by the federal and provincial governments more than education. Provincial governments jealously protect education within their jurisdiction, so the federal government expected British Columbia to pay for and look after the education of its internees. But the B.C. government wasn't interested in educating people who were now thought of as enemy aliens. By the fall of '43, Vancouver schools were closed to any Japanese child remaining in the area. But for the 2,348 school-age children in the camps, there were only two *Nisei* who actually had teaching certificates.

The B.C. government left education up to grade eight in the hands of the evacuees. High school graduates and university students were recruited as volunteer teachers when the first classes were opened in April 1943 in Bay Farm, half a mile from Slocan City. High school children were to fend for themselves. In the summer of that year, 125 *Nisei* took a four-week course given by the education faculty of the University of British Columbia. These students were to be the nucleus of the teachers in the camps.

Propagandists in British Columbia had spread their message well. No welcome elsewhere in Canada was extended to Japanese who wanted to leave that province, and no province was more hostile than Alberta. Japanese workers, recruited by labour-short sugar beet farmers, encountered vocal rejection by citizens of Edmonton, Calgary and Lethbridge. And the halls of academia were no more immune to the attitudes of society. UBC's rejection of *Nisei* students was matched by refusals to enroll them by the University of Toronto, Queen's and McGill. Canadians across Canada failed miserably to uphold their country's grand ideals.

WAR AND THE SUZUKIS

It had been clear for months before Pearl Harbor that Japanese-American bellicosity was approaching an ignition level. Dad sold his car in anticipation of a war. The day after Pearl Harbor, dad went to the barber and told him to cut off his hair in a crewcut. "They're going to treat us like Japs," he said, "so I may as well look like one." Like a samurai warrior cutting off his topknot, it was a symbolic act to declare openly who he was. Dad has worn a crewcut ever since.

The morning after Pearl Harbor, dad went straight to the bank and withdrew his entire account of a few hundred dollars. A few hours later, all Japanese bank accounts were frozen. Dad and mom owned a lot with a small house in Marpole that they rented for income. All their property was confiscated and placed with the Custodian of Alien Property. The lot was eventually sold, and my parents received a statement showing that the money from the sale had all been spent on Security Commission charges. Mom and dad received nothing. They also had endowment insurance policies with Prudential of America which were to reach term when we children reached twenty years of age. All told, about one thousand dollars had been invested, a considerable nest egg in those days. Dad applied to terminate the policies and requested a rebate on the amount already accumulated. He had to sign forms finally requesting the cash surrender, then received a letter from the insurance company stating that the money had been turned over to the Security Commission. My parents never received word from the Security Commission — the money disappeared.

In one of his enigmatic impulses, dad got it in his head that by showing his good faith he could save his brothers from their imminent removal from the coast. Dad went to the Security Commission and volunteered to leave with one of the first contingents of men, on condition that his brothers be allowed to finish the boats they were building. It was agreed to, and because he volunteered, he was given a choice of destination. He chose Solsqua because a number of men from Marpole were going there. At Solsqua, the men worked on building the Trans-Canada Highway. In April 1942, he left Vancouver and his family.

It was mom who had to pack up the house and family when the general evacuation was to begin. Each adult was allowed 150 pounds of luggage while every child under twelve years of age was allotted seventy-five pounds. This had to include everything we wanted — clothing, cooking utensils, furniture, etc. Our prized dolls and toys were left with neighbours and friends, and over the years were forgotten.

In June, 1942, when I was six years old, mother, my two sisters and I left by train with the first contingent of evacuees for the interior of British Columbia. I was so unaware of what was going on that I didn't even realize that all of the people on the train were Japanese. The end of the adventure was Slocan City, an empty town whose ghosts brought back memories of another era.

In the late 1800s, prospectors had scoured the mountains around Slocan Lake in search of gold, silver, lead and zinc. In its heyday in the 1890s, Slocan City boasted hotels, gambling casinos and dance halls. As gold fever subsided people moved away, so that by the 1940s the buildings still standing were decaying relics of the glory years. At the south end of Slocan Lake, the "city" was now a whistlestop for a train, with a dirt road linking it to the outside world.

We were among the first contingent to arrive in Slocan City and got to live in the hotel closest to the lake. We had a small room on the second floor at the back of the building. It must have been a grand building in its day — a large porch ran all the way around it, while columns supported a similar porch above it. But the boards of the porches were so weathered and rotten that we weren't allowed to run around on them. The later arrivals found the buildings fully occupied and so had to spend a winter in tents. Shacks were gradually thrown together in farmers' fields near Slocan that became known as Bay Farm, Popoff and Lemon Creek.

Unlike their Japanese-American counterparts, Canadian Japanese were not fenced in by barbed wire and armed guards. Prison walls weren't necessary in the remote valleys between British Columbia's mountains. There were requirements of travel permits and ID cards, but essentially evacuees had little hope of escape because we were marked by our physical appearance. A small detachment of RCMP in Slocan City enforced the regulations imposed by the Security Commission.

Our building was filthy and cramped. Women shared the cooking facilities and there were large communal outhouses. We bathed in two segregated *ofuro* which were so large I actually learned to swim in one. As children, we weren't aware of the discomfort, the crowding, the deprivation. Instead, we saw it as a continuation of our adventure on the trains. We quickly took it for granted that in the morning we would wake up covered in bedbug bites. But for the grownups and older children, the entire sequence of events must have been a terrible shock. I cannot imagine the anxiety, frustration, fear and anger my mother must have felt in the early weeks. We and thousands of other people poured into the valley with meagre possessions, ill-prepared for the hard winters.

My father in the meantime was living with more that one hundred other men in two bunkhouses in a road camp several hundred miles from Slocan in the Revelstoke Valley. Guards patrolled the camp with guns. Clearing the way for the highway through the mountains, the men were the latest version of Oriental coolies. By hand, the men drilled holes for dynamite into rocks, but the Japanese men were not allowed to handle the dynamite. Dad was paid twenty-five cents an hour. From those wages, $18.50 a month was deducted to be sent to my mother. Room and board at the camp were charged against what was left. There was a lot of rain in that area and the men missed many days of work. Consequently, for the year my dad worked in the camp, he never received any money left over after deductions.

Letters sent out of camp went to Vancouver where they were censored before being forwarded. Sometimes letters took weeks to reach their destination. Dad got the idea of sending magazines to mom with letters slipped inside. They got through uncensored and within days! Once the word got out, everyone started sending out magazines. Of course, someone finally figured out what was going on and it was stopped.

Dad was the only one in the camp who had any experience with horses. Thus he was assigned the job of working with them to skid logs out of the bush. It was a tough job, but he was given no extra pay for it. Once trees were felled and the branches lopped off, a "swamper" would hook up the log with chains. Dad's job was to use a team of horses to pull the logs to a loading area. It required a lot of skill and savvy, and dad was good at it. One of the dangerous parts involved taking logs down a hill because they often started to slide rapidly and the horses had to be kept ahead to keep from being run over. One day, a log began to fishtail and swung sideways down the hill. Dad got the horses out of the way and dove into a depression as the log came down. It rolled over his leg, injuring him and knocking him out of commission. He couldn't work and hobbled around with a cane.

It had been four months since he had seen his family. One day, while convalescing, dad asked for a pass to spend a day in the nearest town. He got along well with the supervisors of the camp, and since he had a bad leg they didn't worry about giving him time off. When he got the pass, he saw that it didn't indicate *where* or *when* he was to go. It merely stated that he had permission to leave the camp. On the strength of that, he decided to chance a visit to us in Slocan City. It was a long and hazardous trip because if he got caught he would have to serve an indefinite detention in Vancouver. He caught a train to Revelstoke where he had to stay overnight in a hotel. He remained in his room the entire time until just before the next train and caught it for Arrowhead.

Arrowhead was located at the top of Arrow Lake and there he caught a stern-wheeler ferry to Nakusp. Dad hadn't eaten at all since leaving camp out of fear that he might attract attention and get caught. So when he got on the ferry, he was ravenous and began looking for the galley. As he approached the dining area, his heart sank when he saw that the cook was Chinese. The cook spotted him. Dad figured he would be turned in.

The Chinese in Vancouver had taken action to distance themselves from the Japanese. There had been a long history of enmity between Japan and China, and Japan's invasion of Manchuria had whipped up Chinese fury on both sides of the ocean. In Canada, some Chinese wore signs declaring, "I'm not a Jap! I'm Chinese." So dad expected the worst. The cook recognized that dad was Japanese and said, "This war not you and me. You and me, same people." He fed dad a big meal and made up a bag lunch. He sent him on his way without taking any money. Dad had always had a great affection for the Chinese who were good customers of his before the war. After this experience, he felt especially close to them. Years later, when I was a teenager, he conceded that if I couldn't find a suitable Japanese-Canadian wife, a *Chinese*-Canadian girl would do.

The ferry arrived at Nakusp where he had to stay overnight again before catching the bus to Slocan. As he walked toward the hotel, two Mounties ambled towards him and once again he was sure the trip was over. Instead of bolting across the street to avoid them, he limped along with his cane and bag, tipped his hat and said, "How do you do?" The Mounties replied in kind and kept going.

After three days, dad finally arrived in the remote town of

Slocan. He had no idea where to look for us. When he got off the bus, he thought a playground would be a place to find us. So he listened carefully and soon heard children's voices. He followed the sounds to some swings, and sure enough, there were my sisters and I. He called us by name, and we looked up and stared. Here was a thin, dark stranger with a cane and a satchel. I thought he might be a doctor, but as he continued to talk, we recognized his voice. We ran to our father, calling out in delight.

Mom had been hired to work as a secretary in the office of the Security Commission soon after her arrival in Slocan. When dad showed up on the flimsy excuse of the ambiguity of his day pass, she was terrified that he would be sent to a detention camp. So the next day, when she went to work, she blurted out the story to one of the Mounties. He immediately summoned dad into the office. The officer informed dad that if the incident was reported, dad would end up in a compound back in Vancouver. "I haven't seen you," the Mountie told dad, "so get on the next bus back to Revelstoke." After only one day and night with us, dad reluctantly retraced his steps back to the camp.

To hear dad tell it, life in the road camp definitely had its plus side. I'm sure life for the young single men must have been hell being so far away from women and some social activity, but dad was married, and because he was an avid outdoorsman he found much to keep himself occupied. He put in a lot of time fishing. It must have been an angler's heaven because the creeks were jumping with trout and the scenery was spectacular. Once the men found a fawn that had probably been orphaned by a poacher and brought it back to the camp. They fed it milk from a bottle using a condom as the nipple, so it was called "Frenchie." Because dad was injured, he was confined to the camp and ended up taking care of the fawn. It became very attached to dad and would follow him around like a dog. It slept under dad's bed and would follow him into the washroom, often nudging him from behind in search of a nipple.

Two of dad's brothers had ended up in the Kootenay Valley east of Slocan in a town called Kaslo. They set to work using their skills as boatbuilders to restore houses for Japanese evacuees to settle in. They did a very good job, and became so highly valued by the Security Commission that when they made a humanitarian request that their brother, my dad, be released from the road camp to join his family in Slocan, it was granted. In the spring of 1943, about a year after leaving Vancouver for the road camp, dad joined us in Slocan.

LIFE IN SLOCAN

When the first classes were opened at Pine Crescent School in Bay Farm, it was April of 1943 and I was seven years old. Our teachers were girls barely out of high school, ill-prepared to handle the job. We walked the half-mile to Bay Farm and I began classes in grade one at age seven. In the fall, I found myself quickly advancing through grades one, two, then three. (A few years ago, I met my grade one teacher whom I had known as Miss Sato. She told me that I had been simply too far along to stay in her class. She recommended that I be skipped and Reverend Tsuji, the Buddhist priest who was principal of the school, agreed.) The only way I could have been more advanced in my education than the others was through having spent all of that time pumping my father's memory of *The Book of Knowledge*. The fact that English was my first and only language — unlike the other students who were mostly *Nisei* — contributed to my doing well in school.

Once I came home from school and told dad that our teacher had said that the Amazon River is so vast that for two hundred yards out to sea the water is fresh. Dad scoffed at that and said she must have meant two hundred *miles*. So the next day, I raised my hand in class and told the teacher, "My dad says the Amazon flows two hundred miles out to sea." The teacher told him later that she was humiliated.

I didn't understand why I was being moved from class to class. In one school year, I went from grade one to being passed into four. But the consequence was that I had no idea what was going on in the classes. I had barely learned to add when the class was suddenly doing multiplication and division. No one bothered to explain to me what we were doing. Teachers simply assumed I knew, because I had been bumped up. Invariably each night I would end up weeping as my father drilled me on my multiplication tables. It was horrifying to me, not because the memorizing was hard, but because I didn't understand why I was doing it and that was frustrating. It was a good thing to remember when I became a teacher, but it was a hard lesson to learn.

Somehow, those years at Bay Farm school didn't turn me off. I think one of the most crucial factors in keeping me interested was my father's incessant queries and interest in what I did in school each day. When he came home from work, he would ask what I had learned and always listened carefully to my responses. It gave me a sense that what I was reciting was important and I loved dredging up all the details. Even today, when something interesting or exciting happens to me, I want to share it with someone right away and I'm sure that goes back to my father's interest. Dad was always critical, informative and extremely helpful.

I don't remember many other children from Slocan days, but I do have vivid memories of Joy Nakayama (now the well-known author, Joy Kogawa) and David Toguri (now an internationally known dancer). Once, some boys mischievously locked me in a classroom with Joy. I felt humiliated at being trapped with a *girl* and frantically dived out a window. David was part of a group of boys dad would take fishing, but even then he was clearly different from us. He often had to miss outings because of his music lessons, something almost unheard of in the camps.

If we could ignore the awesome psychological impact of evacuation, incarceration and loss of property, life for us in Slocan was tolerable. We were buffered by our remoteness from the racism of a society at war with Japan. There was no pressure on the camp dwellers to put in long hours of work to get ahead. In the summer, everyone tended their own small gardens of vegetables. For an avid fisherman like dad, the Slocan Valley was paradise. He got a job working in a store in Popoff. After work and on weekends, he could investigate the creeks, rivers and parts of the lake. As an only son, I was dad's constant companion on fishing and camping trips. If my formal education was suffering, there was plenty to pick up outdoors. Dad's insatiable curiosity allowed him to meet a number of interesting mountain people, prospectors from an earlier time.

Just as with Marpole, memories of Slocan are filled with fishing trips and hikes to gather flowers and stones in the mountains. All Japanese were forbidden to fish, but that didn't stop dad and me. Once he had joined us in Slocan, dad rigged up my own fishing outfit. In an area teeming with fish, to be prohibited from catching and eating a staple of our diet was cruel. But by and large, the RCMP were very humane and turned their backs on our illegal fishing. During the summers, I spent hours on my stomach fishing from the dock where scows and trains were loaded. Once I noticed a line wavering under a scow. Before I could hook it with my line, one of the bigger boys snagged it and pulled in a large trout. I was heartbroken at the missed chance.

On rare occasions, I would catch a trout that wandered in under the docks. But mainly we caught chub and squawfish, a bony, inedible member of the minnow family. One morning, I went down to the dock very early when the dew was still heavy on the grass and the mist was coming off the water. I quietly mounted a platform on the dock and peered into the water to witness an incredible sight. As far as I could see, huge squawfish were milling around the dock. When I dropped in my worm, the water literally boiled with fish trying to hit it. I pulled out fish after fish until I ran out of bait. It was one of the greatest days of fishing I'd ever had. Another time, I was walking along a small creek with a bigger boy when I noticed what looked like a bright red rag in the water. I pointed it out to my friend and he jumped into the water, and with his bare hands, pulled out a brilliant red kokanee, a kind of land-locked salmon.

Dad had all of the creeks and rivers in the area checked out within weeks of his arrival. On occasion, he would get off work early and tell me to meet him after school at one of our favourite spots. Once I was on the way to a rendezvous and came across a black bear blocking the path. I remembered dad's instructions not to panic or make sudden moves. So I just stood there and said, as bravely as I could, "Go away, bear, I'm going to see my dad." And the bear left.

Our favourite spot was a wonderful pool in the creek. It lay beneath a huge rock that formed a deep grotto. We could walk along the edge of the cave and then wade right into it towards the head of the pool. We always caught trout there; what amazed me was that they were almost black. It was a wonderful illustration of the power of adaptation and camouflage, because the fish were almost invisible in the shadow of the cave. I soon learned that early in the morning, when it is still cool, you can pick grasshoppers off plants with no effort because they are too chilled to jump away. They make terrific bait. We also used caddisfly larvae, those amazing insects that glue bits of wood, sand and other debris around them to form a little house. By pulling on the head, we could remove the larvae from their cases and thread them on a hook — a lethal bait. In winter, we would collect golden rod bolls, which are swellings on the plant where insect larvae are sequestered. The worms are small, but they make excellent bait, especially when you don't have anything else. I didn't realize it at the time, but all of this provided me with a base of intimate experience and knowledge on which my scientific career in biology would be built years later.

Each winter, a remarkable phenomenon happened. The whitefish came in to the shore of Slocan Lake to spawn. The snow was usually deep on the ground. Looking out from the dock, as far as you could see in the crystal-clear water, the lake bottom was carpeted with whitefish. Occasionally, you would see a silver flash as one of them flipped on its side to expel milt or eggs. It was spectacular.

Whitefish are terrific to eat and their firm white flesh is reminiscent of red snapper. Dad had a bright idea. He tied a sinker on a line with a large treble hook. He attached a piece of white cloth to the hook so we could see it in the water. Then he would cast it out, let it sink to the bottom, and as he retrieved it, jerk it hard every few pulls. The fish were so thick, this action would often snag one which would then be hauled in and dumped on the snow. My mother loved *sashimi* — raw fish — and in spite of the risk of parasites from freshwater fish, ate a lot of whitefish *sashimi*.

Most people were too timorous to fish so brazenly in full view of the village and they resented us for doing it. Sure enough, one day someone reported dad. I was the lookout, but was so excited by the action that I had totally forgotten to keep a watch. Suddenly, a Mountie appeared right behind us. As he walked us to our apartment, he admired dad's spinning reel, one of the very first he'd ever seen. Dad had an expensive fishing outfit and this man was very reluctant to confiscate it. "Look," he said once we were in the room, "I have to turn in something. So why don't you rig up anything and I'll take that with me." It was an unbelievably generous and humane thing to do. Dad fixed up some line with a sinker and hook tied to a stick and that's what was turned in. I have had a soft spot for the RCMP ever since.

A world war was raging, and for everyone in Canada, there was rationing and sacrifice. But we were considered enemy aliens, and fresh food was even less easy to come by in remote areas like Slocan. The people who saved us from malnutrition were the Doukhobors. They farmed the area, and came into town with wagonloads of fresh vegetables and meat. They had a captive, albeit poor, market. My father always speaks gratefully of the Doukhobors because they were friends of the Japanese-Canadians.

In Slocan, a friend of mine told me he had learned a swear word in the Doukhobor language and taught it to me. When a Doukhobor farmer came along with his wagon, we leaned out of a second-floor window and began to chant this word. At first he ignored us, but as we persisted, he finally seized a huge machete, jumped off the wagon and ran cursing into our building.

We were terrified! We fled into another apartment and dived under the bed. We shivered in fear under that shelter for what seemed like an eternity.

Today, Valhalla Park occupies all of the western ridge of Slocan Lake. When we lived there, dad and I would row about two miles up the lake to Evans Creek. We'd troll on the way and usually catch some nice Kamloops trout (rainbows). Once we went over in mid-June. We fished up Evans Creek to Evans Lake and then hiked from there to Cahill Lake where there was a log cabin. It was a tough climb of several miles with lots of windfalls to climb over or under and devil's club plants to sting us. It was night when we arrived at Cahill. I was tired and immediately lost consciousness in my sleeping bag.

I woke up in the morning to a rasping sound in the cabin. It was a porcupine probably attracted by residual salt in some of the wooden implements we had handled. Dad grabbed a sweater and struck the animal on the back. It leapt about trying to get away. I pleaded with dad to let the animal go because I thought he was trying to hurt it. But he was only after the quills that remained stuck in the sweater, because they make excellent floats for fishing. As the porcupine trundled off outside, I had learned what many people still don't know. Porcupines *do not shoot* their quills. They easily come loose on contact.

We continued hiking to Beatrice Lake, which was so high up that we encountered snow. We found a makeshift raft at the lake, climbed on and started to fish. We caught many. They were all stunted, about ten to twelve inches long, and looked like eels because they were so thin. We fished to the other end of the lake, and on the way back a storm blew up. I didn't realize how close we were to real danger. We couldn't put in to shore because the banks were rocky and steep, and as we blew down the lake, the wave action broke the raft up. It disintegrated completely just as we reached shore. That night, dad built a lean-to with logs and pine boughs and heaped wood on a huge fire which he kept burning all night. With the wind and snow blowing all around, we were as warm as could be. It was an adventure of a lifetime!

Slocan Lake was great for swimming as well as fishing. After

I learned to swim underwater in the *ofuro,* I would go down to the shore and in the hip-deep water dive under and swim until I ran out of air. Then I would stand up. One day I tried to reach a barge. As I walked out, I gradually sank in up to my neck. Standing on tiptoes, I was still about fifteen feet away from the barge. So I gingerly turned around to go back to shore. But the bottom sloped rather sharply and sand began to slide out from under my toes. I found myself slipping into deeper water. In desperation, I ducked my head underwater and started to swim toward shore. Running out of air, I tried standing, but still no bottom. I kept bobbing up and down, kicking off the bottom to get my head out of water to gulp some air, swimming a bit underwater then kicking off the bottom again. I could have drowned, but instead I learned to swim!

One day I was sitting on the porch of the hotel that was our home when people started running down to the lake. I didn't know what was going on, but followed the crowd down to the shore in time to see some men pulling a white body from the water. A mother was wailing, "Takebo! Takebo!", the drowned boy's name. He was covered with a towel and I pushed in close for a look. I had never seen a dead body before and was struck by the whiteness of his skin and the incongruity of a living grasshopper walking about on his lifeless leg.

Most kids our age in the camp were still *Nisei* like my parents. So, like my parents, they were also fluently bilingual. Many went to Japanese school to keep up the language. My father didn't believe this was necessary for his Canadian children. English was the only language used in our home, so my sisters and I didn't understand Japanese. This created a lot of tension between me and the other youngsters: my inability to speak Japanese made me an outsider. I had a few friends in Slocan, but none that was close. It was all the more lonely for me because the few children of the Doukhobor farmers in the area wouldn't have anything to do with me either.

My peers sometimes echoed their parents' bitterness and hoped Japan would kick the hell out of the Allies. I didn't really know what war meant, but was always loudly and fiercely Canadian.

As a result, I was often beaten up at school. My father and mother always encouraged me to be outspoken. Among the more reticent *Nisei,* this kind of brashness was resented.

THE JAPANESE

On the whole, the Japanese of British Columbia obediently carried out the succession of orders that stripped them of even their second-class rights and their faith in Canada. There were those who counselled resistance, *gambari,* but the majority sentiment was to prove that Japanese were good citizens by living up to all of the demands put on them.

The final indignity was the refusal by the B.C. government to allow any Japanese to stay in the province after the war. For years there had been pressure within British Columbia to eliminate the Yellow Peril by sending Asians back to the Orient. The problem was that a lot of them were Canadian citizens by birth. War and evacuation provided a way of solving the "Japanese Problem" in the short run. The goal was articulated in the nomination speech by Ian MacKenzie, the Minister of Pensions and National Health, on September 18, 1944: "Let our slogan be for British Columbia: 'No Japs from the Rockies to the seas.' "

In the same year, Prime Minister Mackenzie King announced his plans for the Japanese after the war — deportation and dispersion. He would set up a commission to examine the background and loyalties of all Japanese. Those volunteering to go to Japan, or judged disloyal, would be deported. Those pronounced loyal would be dispersed across Canada. The Canadian policy was a striking contrast to the United States. Because of the favourable turn in the Pacific War Zone, the American government decreed in December 1944 that evacuees could return to the coast. In Canada, early in 1945, Labour Minister Mitchell announced that all Japanese over the age of sixteen had to indicate to the RCMP whether they intended to "repatriate" to Japan. The dilemma for internees was stark. Mitchell's promise was this: "Free passage will be guaranteed by the Canadian government to all repatriates being sent to Japan, and all their descendants who accompany them, and including free transportation of such of their personal property as they take with them."

For those who chose to remain in their homeland, adopted or native, the other side of the solution was clear: "Japanese Canadians who want to remain in Canada should now reestablish themselves east of the Rockies as best evidence of their intentions to cooperate with the government policy of dispersal. Failure to accept employment east of the Rockies may be regarded at a later date as lack of cooperation with the Canadian government in carrying out the policy of dispersal."

All of this was being invoked even while the Prime Minister admitted: "It is a fact that no person of Japanese race born in Canada has been charged with any act of sabotage or disloyalty during the years of war."

The Japanese-Canadians were hard-working, honest people who had committed no crime other than sharing genes with an enemy of Canada. Now they were bitter, disillusioned and angry — and had every right to be. At the time the offer to repatriate was made, Japan still seemed capable of winning the war. Some people were dazzled by the hope that they might get good jobs in a victorious Japan. Others wanted to register their disgust by rejecting Canada, the country that had treated them so badly. The Japanese spoke of *gambari* — resistance — and thought of signing to repatriate as a way of protest and of maintaining pride.

Japanese people tend to act collectively. Within the internment camps, people had survived the hardships of being uprooted, having their property sold off at bargain-basement prices, their bank accounts frozen and every right of citizenship suspended. Not surprisingly, there was strong sentiment within the camps to sign up and go to Japan, even though that meant going to a totally foreign country for many. In the camps, 81 percent of all Japanese over the age of sixteen signed up to repatriate. This was a striking contrast to only 15 percent of those living east of the Rockies.

The pressures to sign up to repatriate during the war were strong. Dad and mom were the *only* people in Slocan who did not sign and they paid the price in social ostracism and contempt.

My mother belonged to a group of women who met regularly to gossip and exchange news. When mom chose to stay in Canada, one of the women in the group called her a *saibashi* (a derogatory term for native Indian) and said mom had no right to belong to the group. Mom was deeply wounded and never went back. Although she told dad of the incident, she went to her grave without identifying the woman.

One night, my father came home discouraged by all the talk and pressure to sign up. "What would you think about going to Japan?" he asked me. I hurled myself onto my bed, crying uncontrollably. I had been beaten and taunted by the *Nisei* kids who made fun of my inability to speak Japanese and my pride in being a Canadian. I did not want to go to Japan after all that. That was the only time dad ever wavered about staying in Canada. He never mentioned it again.

In the winter of 1944, I began to resist attending school. I would come home dripping wet, and mope about indoors. Dad was worried and one day came out along the tracks to meet me after school. In the distance, he could see a kid being chased by a pack of boys. The victim would dart off the track into the deep snow and struggle, while the others would set on him like a pack of wolves. The boy was me, forced to run a daily gauntlet because my family had chosen to stay in Canada. Those experiences are seared into my memories and produced an ambivalence towards other Japanese-Canadians. Intellectually, I identify strongly with this minority group, but emotionally, I still feel the pain of the terrible rejection by my own people in camp.

My mother's sister, Aki, and her *Issei* husband signed up to repatriate. She was reluctant to leave her native land for a foreign country, but her husband was adamant, and as a dutiful wife, Aki had no choice. Her three Canadian-born sons had no choice, either. They, too, would go to Japan. My mother's parents also signed up to return to Japan after almost forty years in Canada. Postwar Japan was a bleak place even compared to the conditions of internment. Within a year, both of my grandparents and my uncle were dead.

Dad had hoped to hang in and eventually resettle back in Vancouver after the war. But the B.C. government was determined to get rid of all Japanese. In the spring of 1945, as the war was drawing to a close, the "repats" were being collected in centres to await the boats to Japan. They were treated well, while people like my parents were classed as "undesirables" because they had chosen to remain in Canada. The small number of non-repats were being assembled for dispersal out east.

In the spring of 1945, we were moved to Kaslo to join others awaiting movement out east. We moved there while I was near the end of grade four and we only stayed for a few months. Dad got a good job handling horses to skid logs for a logging camp. The pay was good and he was well liked. He hoped he would be able to stick it out in the Kootenays and remain in British Columbia. But the Security Commission informed the head of the logging company that dad was an "undesirable" because he was remaining in Canada. Dad was fired, while two men who had signed up to repatriate were kept on the payroll.

Kaslo was like a large town compared to Slocan City and had a substantial population of Caucasians. It was in the middle of the day on August 15, 1945 and for some reason, I was having a bath in an *ofuro*. There was only one other person in the tub with me, an old *Issei*. Suddenly, bells started to ring and we heard horns blaring and people shouting. *"Maketa!" (We've lost)* cursed the old man, and I realized that the war must have ended. I dressed quickly and went out onto the streets where people were celebrating and letting off firecrackers. I went up to one of the vendors to buy fireworks. A big white boy came along and kicked me in the backside and yelled, "Go away Jap. We beat you!"

As the battle tide changed and imminent victory over Japan became apparent, people who had signed up to repatriate went to the Security Commission and asked to have their names taken off the list. Many of them soon left for the east.

There is an enduring bitterness among those who, like my father, had declared their intention to stay in Canada and had been castigated for it. They regard those who signed up to repatriate as traitors to Canada. But what was even worse, they were also cowards because they didn't have the courage of their convictions to go through with it. I am much more sympathetic than my father with the plight of the repats. I do think it was terrible for them to turn their backs on Canada, but when you think of what they had gone through, it's hardly surprising. Knowing Japanese-Canadians, I understand the extent of the social pressure to conform and act as a group. Few would be able to resist that. Once they signed up to repatriate, I believe they shouldn't be blamed for backing out at the end. They had made a mistake and it would have been foolish to compound it by going through with it *just to save face*.

Dad's brother, Shuji, and his wife, Kaz, left Kaslo and ventured east long before the end of the war. They had ended up in London, Ontario where they worked for a doctor as a cook and maid. There they no longer encountered the deep hostility so familiar in British Columbia. Soon Shu had a more suitable job building boats. He wrote to his older brothers and urged them to come. Dad always knew the opportunities were better out east, but he wanted desperately to stay in his beloved British Columbia.

When dad lost his job in Kaslo, however, he realized that his hope of remaining in British Columbia was not great. He gave in to the pressure and finally followed the advice he had dispensed to the rest of the family — go east, where people are more tolerant, and start your life over again.

Finally in September 1945, we were exiled from the west coast. We were sent off on a long train journey across Canada to settle in Ontario. There we began all over again in a less bigoted part of the country. Like many of the dispossessed who came to this country at the end of the war, we had a few suitcases of possessions, no money, but a lot of determination. We were starting out at exactly the same point that my grandparents had when they first arrived in Canada, though we had always been Canadians.

My favourite shot of Mom when I was about eight months old.

An expert fisherman at the age of four.

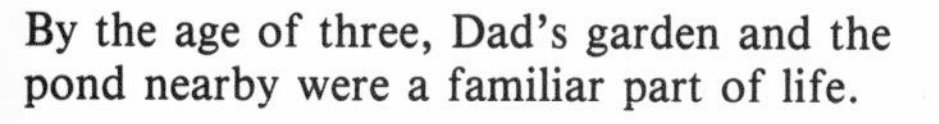

By the age of three, Dad's garden and the pond nearby were a familiar part of life.

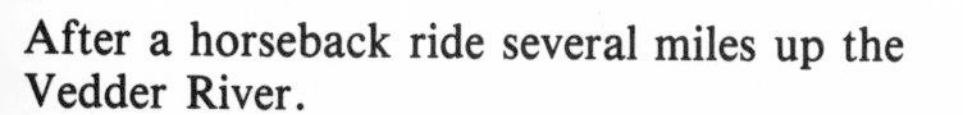

After a horseback ride several miles up the Vedder River.

My first catch at Loon Lake when I was four.

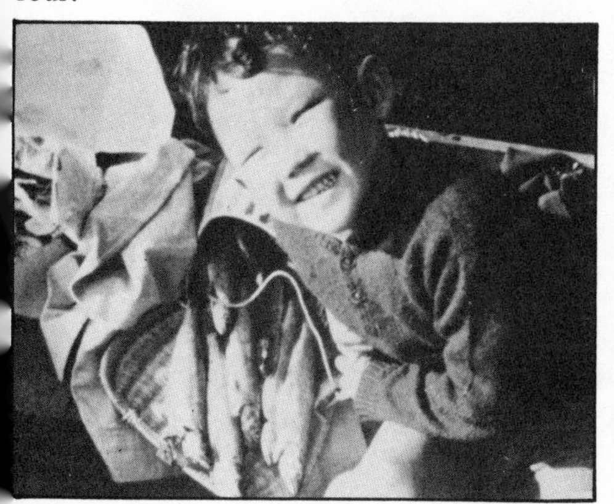

The first day of kindergarten for Marcia and me, September 1941, three months before Pearl Harbor.

Slocan City. We were joined by Dawn, conceived and born in the camp.

The MacGregors, our next-door neighbours in Marpole. That's my best friend Ian at the left in the front row, then me, Aiko and Marcia. At the back are Ian's older brothers, his parents and mine.

CHAPTER THREE

OUT EAST — BEYOND THE MOUNTAINS

LIKE REFUGEES FLEEING A WAR ZONE, we started our journey across the continent in a day coach with our battered suitcases of possessions. We had been expelled from British Columbia — our home — and as we left the foothills for the expanse of the prairies, my parents must have been preoccupied with anxiety and fear. But they buffered us children from the turmoil of constant change. I never questioned why we had to move, where we were going or what we would end up doing. That was for mom and dad to worry about. How trusting we were in their strength and wisdom. We kids just scampered around the train and watched the endless prairies blur by. Finally, the rocky shore of Lake Superior relieved the monotony, and like an immense sea it accompanied our train for hundreds of miles.

When we reached Toronto, we were put up in a hotel in Islington where we stayed for a number of days. Dad found a job in Essex, the southernmost county in Canada. It was on a peach farm in a rural community called Olinda, a few miles from Leamington.

In Olinda, my sisters and I attended a one-room school with eight grades taught by a wonderful woman, Miss Donovan. Now I realize that she had prepared those children for our arrival. At the very first recess, I was shocked to be pulled right into games. Everyone seemed to want to tag me and I felt the centre of attention. In Slocan and Kaslo, I had learned to fear Caucasians because of the abuse I suffered. And here were these children behaving as if we were just like them. It was wonderful! Everyone was aware that we were Japanese, but many of the families in the area were recent immigrants themselves. I encountered nothing but kindness from them all.

My years of internment had left a permanent mark on me. In my younger years in Marpole, I had felt no sense of difference — Ian MacGregor had been my best friend and I was a Canadian. But I didn't know a single Caucasian child for those three years in camp. Slocan City had thrown me together with others solely because of our shared genes. Being one of the few *Sansei* and not speaking Japanese, I didn't fit in with the other Japanese children, nor did I share their knowledge of and affection towards Japan. I undoubtedly aped my outspoken father and was harassed for my strong opinions.

There were only four of us in grade five in Olinda. One was a lovely blonde Lithuanian girl over whom I developed my first crush. Olinda was heaven for a child my age. My chums and I had spent the summer exploring the fields, woods and creeks. I learned how to ride a bicycle and borrowed my dad's bike to hang out with my friends. Our family stayed in a big house belonging to the owners of the peach farm where dad worked the orchard. I have never eaten peaches more juicy and delicious. First to ripen were the Red Havens, small and sweet, then later the large Golden Jubilees. I got to drive the tractor while dad would load the baskets of picked fruit onto the trailer.

Olinda is where I started collecting insects. Dad made a hoop of stiff wire and attached it to a long wooden handle. Mother then sewed mosquito netting onto the frame and *presto!* I had an insect net.

There's no better way to learn about nature than by collecting insects. One quickly learns where to find different kinds. A cowpat is a terrific source of dung beetles; a rotting stump is rich in wood borers. A fabulous area for indiscriminate collecting is on a sandy shore near a lake after a strong wind has washed up all kinds of insects.

Most people start off with butterflies because they're so lovely and easy to spot, but I preferred moths. The light above our back door yielded some of my most prized moths, including a lovely green-white luna! Even now, I can remember when and how I caught the ones which were the most prized. It wasn't their size or beauty so much as how rare they were or how hard they were to catch. I would spend hours arranging the insects so their legs were spread out, the wings laid flat and the antennae perfectly displayed.

My friends were children of farmers and I learned a lot from them. They had an understanding of plants and animals, as well as weather, soil conditions and seasons. In the summer, they always went barefoot. Once they took me inside a greenhouse and dared me to take off my shoes and walk across the sand. I did, and learned what "tenderfoot" meant. The sand was burning hot and I couldn't bear it, while my companions felt no discomfort. That summer I abandoned shoes, and in a few weeks my feet were just as tough and calloused as theirs. I remember going to roast hotdogs in a woods one night. We crashed through the brush, treading painlessly on twigs and stinging nettles. Years later, when I was in the Kalahari Desert for a film on the Kung people, I could see how thick and cracked their feet were — just as mine had been that summer as a kid.

In Ontario, we discovered brand new areas to explore for fish. Dad was soon checking out the creeks and ditches around the farm, and one of the first things we found were ponds in sand quarries that held sunfish. They were exquisite. We would scoop them up, take them home and keep them in glass jars. Once we noticed a dark cloud in a pond that moved slowly through the water. We quietly crept over to discover a thick school of baby catfish. We also caught little gar pike, bowfins, carp and painted turtles. Dad once found a snapping turtle shortly after it had hatched. We kept it in an aquarium for years. At that time, I decided that I wanted to make a living studying fish.

We were constantly foraging for plant and animal life. Dad would catch perch, catfish, sunfish and bass, and mom prepared them in all kinds of ways for our meals. We learned where to gather asparagus along the railroad tracks and made salads of different kinds of weeds. We dug up the roots of burdock plants which we called "gobo." They were delicious and wonderfully crunchy when cooked with soy sauce and sugar. And mushrooms! They grew profusely in the fruit orchards. It was in Olinda that I had my first puffballs and morels fried in butter.

LIFE IN LEAMINGTON

After a year in Olinda, dad got a job working for a dry cleaner in Leamington. We moved into one of the new boss's houses on a half-acre lot. Leamington is the southernmost town in Canada, right above the spit of land that is now Point Pelee National Park. It is also the home of the Heinz Company. We were the first non-Caucasian family to live in Leamington. This was no small thing, as one of the boasts of its citizens was that no coloured person had ever stayed in town overnight.

My grade-six teacher was a woman who was a pillar of the school and community. One day while I was sitting in her class, she suddenly told me to get out of the room and wait for her outside! Confused and humiliated, I slunk out and waited with trepidation. When she came out, I immediately asked her what I had done wrong. "You were smirking at me," she said, "and I know what *you people* think." To this day, I have no idea what she meant, but the term "you people" still sends shivers up my back.

The first summer we were in Leamington, dad's boss made a proposal to my dad. He suggested that they go into a joint farming venture. The boss would carry all of the expenses, provide the land, equipment, fertilizer and other costs, while dad would provide the labour. After the boss got his expenses back, the profits would be split. The crop was to be onions. Dad and mom would oversee the operation, but essentially I, with some help from my sisters, would spend the summer weeding and hoeing the onions. Dad promised that I would have a share in the profits. He held out the possibility that I might make a hundred dollars, enough to buy a brand new bike and still have money left over. So we worked hard. We planted the seedlings, fertilized them and hoed the weeds daily. It took a lot of time, but the onions grew well.

When they matured, we harvested the onions by pulling them and leaving them on the ground to dry out. Unfortunately, it rained before they had completely dried. We topped them, sacked them and sent them off to market. They came back, rejected because of rot that had set in after the rain. We had to dump the onions back out and check them all again. Very few were salvageable. Dad's boss lost all of the money he had invested and we didn't get a thing for our work. At the end of the summer, dad gave me the bad news. Because he knew how hard I'd worked, he gave me ten dollars. I wept in bitter disappointment for hours that night. It was the first and last time I set my heart on something beyond my means.

The great joy of Leamington was Lake Erie. There was a marvellous dock near Seacliff Park to dive off in the summer. Years later, when I went back to visit the dock, I was amazed at the height of the pilings. We dove from them without fear, but I doubt that I could do it now. I spent a lot of time fishing for perch off that dock and would bring home strings of fish for dinner. Perch are one of the finest eating freshwater fish. But what a difference forty years makes. So many toxic pollutants have been dumped or leak into the Great Lakes that the fish are poisonous. Signs posted around Lake Ontario warn people not to eat more than a few ounces of fish a week.

Each spring, an amazing biological phenomenon took place — the hatch of insects called mayflies. They were delicate four-winged flies about three centimetres long. After spending a year in the water in the nymphal stage, at the right moment the mature nymph would rise to the surface to metamorphose. A fly would then emerge from a split in the skin. The adults were essentially flying gonads, geared solely to find a mate, then reproduce and die within a day or two. What was most remarkable about these insects was their numbers. They hatched so copiously that cars skidded on their carcasses on the roads, and they cut down visibility with their sheer mass. They covered the sides of buildings and at night the beating of their wings made a loud noise. On the shore of the lakes, their bodies would wash up in immense piles over a metre deep, throwing the fish into a feeding frenzy. I can remember casting out into the water and pulling out fish called sheephead one after the other. Birds and other animals gorged on these insects and the fly carcasses nourished untold numbers of plants and micro-organisms.

Today, pollution has all but destroyed that great hatch of mayflies. To youngsters who grew up after they disappeared, nothing seems amiss. Indeed, they might even think those hatches must have been messy and smelly. But what an appalling loss to the ecosystem. We shouldn't forget this loss or else we won't heed its warning.

At that time, Point Pelee was not yet famous as a bird sanctuary. We would bicycle down to the swamps and observe the herons and bitterns there. And I loved gathering fossils on the shore. The thought that these remains and imprints in rock were once living creatures long before there were human beings filled me with amazement. But dad and I were really more interested in fish. Dad would cut the heads off the different fish we caught or found on the shores and take them home and dry them out. Then he would shellac them and mount them on wooden plaques. We had a marvellous collection. As well, when we found fresh-killed animals on the highway, we would haul them home and skin them. We would salt the skins and stretch them over boards to dry out. For me, these were wonderful opportunities to learn anatomy.

We gathered worms for fishing by going out on lawns at night with flashlights. This was in the mid-'40s, long before dew-worms were the big business they are now. Dad got the idea of selling them for a penny a piece. He experimented with all kinds of ways to keep them alive and found the best thing was strips of cardboard saturated with water. After a night of collecting, we would come home with hundreds of worms and put them in the refrigerator. Many mornings I found my mother patiently pulling errant worms from the vegetable crisper. She never complained.

In my room in Leamington, the walls were plastered with pictures of fish, animals and birds that I cut out of dad's fishing magazines. I would pick up birds killed by cars and cut off their wings and dry them out to mount on the wall. My bedroom was an amateur's museum. My bookshelves were loaded with fossils and rocks. Dried fish heads were glued on plaques, and snake and animal skins were stretched out on the walls. I kept aquaria with freshwater fish, and boxes of pinned insects.

One year, a Japanese man who worked at the Royal Ontario Museum came to our home. It was a great honour to have such a scholar visiting us, and I was in complete awe. Dad proudly ushered him into my room where our guest made polite sounds of admiration at my collections. "I see you like nature," he said of the obvious. "What do you want to do when you grow up?" I told him I would like to be an ichthyologist — someone who studies fish. "That's a good thing to do as a hobby," he replied, "but choose something practical to make a living." I was shocked and bitterly disappointed. How sad that a man who had what I considered a wonderful job could have regarded it so poorly as a profession. His remarks made a great impression on my parents, but I never found them persuasive. In advising youngsters today, I always tell them to go where their hearts pull them and not to try to figure out the best thing for income or long-term security. I often think of that man as I say it.

When I was eleven, I began to work on farms during the summers. My first job was picking berries for a few cents a basket. Next I went to work on a potato farm for a man's wage of fifty cents an hour. I laboured hard on a potato digger, a big machine made in two parts. The first section had a big blade that dug into the ground. When the tractor pulled it, dirt, rocks and potatoes came up on a big conveyer belt. One person stood on a running board, grabbed the leaves of the plants, and gave them a shake to knock the potatoes off. The potatoes then spilled onto a second conveyer belt that had three people on each side, tossing out dirt, rocks and other debris. At the end of the belt, the potatoes dropped into sacks. A man would lift the sacks off when they were full. It was back-breaking work, and at the end of each row or two everyone would switch places so that no one would get stuck with the worst position.

Once when I was working on a grader that separated potatoes by size, my boot got caught in the revolving mesh. It tore the leather open and yanked my large toenail right off. Needless to say, it hurt but we just wrapped my toe in a hanky and kept on working. It wasn't bravado; it simply didn't occur to us to do otherwise. There was never any thought of seeking compensation or protection. It was just an occupational hazard.

There was a lot of truck farming in Essex county. We worked on different farms or would shift from field to field as crops ripened. One of the crops we harvested was celery. We were given a small knife with a curved blade. We'd pull a plant out of the ground and trim off the leaves and roots. One of the common results of working with celery was "celery rash" wherever our skin was exposed to the leaves. This was a terrible itch accompanied by blisters which left purple scars that lasted for years.

I took pride in each job and tackled it as a challenge, to be done as fast and as well as possible. We had to pick berries quickly because we got paid by the basket. But we would never dream of filling the basket with green or squashed berries. I always felt competitive and tried to be better than anyone else.

There was never any thought in our minds that we Suzuki children would end up as seasonal labourers on farms for the rest of our lives. Education was all-important because that would be our avenue out of poverty. We took whatever jobs we could get, but school was our parents' top priority for us.

I've never forgotten those years working as a farm labourer — the difficult conditions, the lack of health protection or any financial compensation for accidents. When we buy our meat and vegetables conveniently and attractively packaged, we should be aware of the human cost involved. It appalls me that today in Canada we still exploit migrant workers who are completely locked into a cycle of poverty and total vulnerability.

In those years in Leamington, I suckered flu tobacco and picked leaves of burley tobacco plants. I harvested lettuce, cucumber, beans and peas. And, oh, those tomatoes! I still have vivid memories of those incredible tomatoes. Most of them were bound for Heinz, but many were also shipped out to markets. They had a powerful smell, and when you squeezed them the seeds and juice would come squirting through the thin skin. When you bit into them, they were as sweet as peaches. I loved it when my mother made tomato sandwiches. When I peeled off the wax paper, the bread had sopped up the juice and I could practically drink the sandwich.

In 1968, the International Congress of Genetics was held in Tokyo and I attended as a Canadian delegate. After the conference, a few of us were invited to Mishima where the National Institute of Genetics is located. We were shown the facilities and invited for lunch. At the end of our lunch, two watermelons were brought out for dessert. Remember, we were all geneticists, so here was a chance for our hosts to show off some genetic strains. The first fruit was enormous, a dark green American watermelon that was almost a metre long. When our host cut it in two, he pointed out that the centre was the sweetest part with about 20 percent sugar content, which fell off steadily to about 14 percent at the outer edge. That's why the centre is always the sweetest.

Then a Japanese watermelon was brought out with a great flourish. Its outside was beautifully marked, a light green with lovely darker bands. But the big surprise was that it was tiny, a third of the size of the American breed. When it was cut open, the host pointed out that sugar content was 20 percent in the centre and stayed at 20 percent all the way to the rind. That really struck me. To the Japanese, a meal must feed the eyes as well as the stomach, and the quality of *flavour* matters far more than mere quantity.

In our society, we have paid an enormous price for the abundance and uniformity of our food. We lack diversity in appearance and flavour, a reflection of the fact that we are mass producing vast acreages of genetically identical crops. Usually, the plants are fashioned for qualities that render them dependent on a highly controlled chemical regimen of nutrition and pest control, while permitting them to be handled with machinery. Rapid growth, size, storage qualities — these are important traits sought by the breeder to maximize profit. There's nothing necessarily wrong with that — it's just that this is often accompanied by loss of those qualities of variability, flavour and aroma that make eating such a pleasure. We urban consumers have grown used to having fresh fruit, vegetables and meat year round and seem prepared to pay the price in palatability.

In the fall of 1949, I started grade nine at Leamington High School. There was only one high school in the Leamington district, so not only did the kids from town attend but students were also bussed in from neighbouring farms. Many of the students were farmers' children and knew about hard work. Few had a lot of time to hang around the restaurants or pool halls. There were several sections of grade-nine students, and I was in the section for the youngest students. My class was considered to be bright, because a number of us had skipped a grade.

Leamington High School was a small school with a great deal of teacher and student school spirit. Teachers devoted a lot of time to extracurricular activities with their students. All incoming students were divided into one of four "houses" — Alpha, Beta, Gamma or Delta. Once we were assigned to one of these houses, we belonged to that group for the rest of our high school careers there. Students in the upper grades made a point of meeting newcomers in their houses, so there was an early mixing of different grade levels. Points acquired in student competitions determined the standings of the houses. I enjoyed the competition and the sense of belonging. I felt accepted. I respected most of my teachers. In short, I loved my new high school.

One of the major events at Leamington High School was a school-wide oratorical contest. There were four categories: junior and senior classes for boys and the same for girls. Each house held its own competition for contestants to be entered in the finals. Teachers and experienced students served as advisors and judges. It was a terrific experience. My father had always emphasized the need to be able to speak extemporaneously. He had always felt embarrassed at how poorly many Japanese-Canadians spoke in public. I decided to try out for the oratorical contest and surprisingly, at least to me, I was chosen as one of the finalists in the junior boys class representing Gamma House.

The finale was quite an elaborate affair. Contestants in each category competed in front of the entire student body. Each orator had to give a five-minute prepared talk. My speech was taken from an article I found in *Reader's Digest*. It was a description of the incredible wonders of the natural world. I remember three of them: Anableps, the fish with four eyes that swims with two above water and two below; the remarkable snail with a rasping tongue that has thousands of teeth on it; and the water ousel, a bird that literally flies underwater. It was a good choice, full of fascinating information and vivid images, exactly the same ingredients of a good television report.

During the contest, each contestant drew a topic from a bowl and had to make up a three-minute extemporaneous talk. The topic I drew from the bowl was "If I Were Three Inches Tall." I made up a story about being three inches tall and going fishing one day with my father. I fell out of the boat into the water and was eaten by a big bass. I used my knowledge of fish anatomy to describe the path I took, slithering down its digestive system and using my penknife to burrow into its air bladder for air to breathe. Suddenly, everything seemed to be turning upside down and I could feel the fish being pulled upward. Then, miraculously, light appeared through a hole that was being cut by a knife. It was my father who had caught the bass and in cleaning it had rescued me. My speech got a lot of laughs and no one questioned how a distraught father who had just lost his son overboard would still be fishing. But more important, I won!

My happiness in grade nine was interrupted early in the school year when my parents decided to move to London, Ontario. When I begged them to let me finish my year in Leamington, they arranged for me to stay with the Shikaze family, farmers who lived a few miles outside of town.

Both of the Shikaze parents were *Issei* who had emigrated to Canada as adults, so they spoke very little English. I learned what little Japanese I know today through total immersion with the Shikazes. They had two sons who were just a few years younger than I. The boys were seasoned farmers who worked like Trojans and put me to shame. They treated me well and taught me a lot about hard work and obedience. In return for my room and board, I worked early in the mornings before the school bus arrived, then after school and on weekends. I worked on the farm through the summer of 1950 then said goodbye to the Shikazes to join my family in London where they had moved almost a year before.

THE BIG CITY — LONDON

When I arrived, I found that my sisters already were well ensconced in school with friends. They seemed to have made the transition to school in London far more easily than I thought I would. Because I had skipped two grades, Marcia was behind me, and she, Aiko and Dawn started off in elementary school.

Dad had moved to London at the invitation of his brothers who had established a successful construction business called Suzuki Brothers Construction. They specialized in single-family homes and were known for the quality of their houses. Dad joined the company, not as one of the partners, but to work as a kitchen cabinetmaker. It was difficult, as the eldest in the family, for him to come in as an employee, but he saw it as a way of making more money for our family and he felt the schools in London would be better for us.

In many ways, dad had been a disappointment to his parents. He had always seemed to indulge in what they regarded as frivolous interests — fishing, camping, gardening, even drinking — impractical distractions from the main business of making money and gaining security. This was in contrast to his brothers who were good providers, teetotalers and regular churchgoers. My uncles had a thriving business and were affluent by our family's standards.

I was fourteen years old when I arrived in London. I had an older cousin, Dan, who was a year ahead of me in school, and his brother, Art, was a year behind me. We three and my sisters were the eldest grandchildren in the Suzuki family. There were fourteen others who were several years younger. I felt distant from my cousins because of their relative affluence and assimilation. My cousins were completely integrated. It was not an accident that of all the Suzuki grandchildren, who now number twenty, only my twin sister and I married Japanese-Canadians. We were all *Sansei,* but unlike my sisters and me, none of our cousins had stayed in camps in British Columbia. Soon after evacuation, they had moved to Ontario, where they had been readily accepted by their neighbours and schoolmates.

Periodically, the entire clan of Suzukis would converge on my grandparents' farm at the edge of town. I remember the farmhouse had a peculiar smell of pickled cabbages and soy. My grandparents, who shuffled about in a strange mélange of Western and Asian clothes, had six sons and one daughter. With their mates and offspring, family gatherings were potential bedlam. In good weather, the farm was great because there were acres of fields and woods where we could roam. A creek was alive with frogs and turtles, and we would flush rabbits and pheasants quite often.

But in the winter, we would all remain inside the house. The women would do the cooking and care for the babies, while the men would talk about business and worldly matters. There was no television then, so the older kids would listen to the discussions of the elders. None of us, however, was ever expected to participate, only listen.

The thing I remember most about our gatherings were the prayers at the beginning of the meal. We were a classic picture of a multi-generation immigrant family. The grown-ups would be at a big table with a few of us older kids sprinkled among them, while the little kids would sit at a separate table within short reach of their mothers. My grandfather would start off a Christian prayer in Japanese, *"Tenno kamisama"* (Revered God), pause, take an audible suck of wind, then another short phrase, pause, suck, and so on. Grandfather's prayers seemed endless, especially since we children had no idea what he was saying. We would begin to snicker and shoot glances at each other, trying to induce someone to laugh. The youngest kids would soon be gurgling and moving around, while the older ones would receive a sharp swat on the head from a parent. In retrospect, I think my grandfather was talking more to my parents and uncles and aunts than to God. The whole ceremony was very ritualized along Japanese customs of filial piety. But we children were Canadians and represented the dead end of that cultural tradition.

One of the first times I visited the farm, I heard a bloodcurdling scream in the house. It was my young cousin who, with his parents, was staying with our grandparents at the time. He had a stomach ache and my grandparents were treating him with *okyu*. This was a bizarre ritual in which a small bit of material resembling punk would be placed at an appropriate spot on the skin and lit with a match. It would burn very slowly and must have hurt like hell as it got near the skin. This ritual enhanced my sense of alienation from my grandparents.

My father has scars all over his body from having had *okyu* applied to him. During his childhood, he had *okyu* burned on his head, neck, arms, stomach, back and legs. And he still continued to have it done to him while we were growing up. He would direct my mother to apply the wad at different places, and as it burned down, he would groan and say, "Owww, that was a big one!" Dad always claimed the heat stimulated circulation near the skin, but my own interpretation is that once the burning stopped, any previous pain would have paled in comparison. I was grateful that my parents were Canadianized enough not to impose the pain on me.

At the farm, I often listened to my grandparents berating my father or exhorting him to be more serious like his brothers. Once I peeked in to see my father weeping in front of his mother. I was furious to see him treated like that and thereafter would always kick up a fuss about having to visit the farm. Dad interpreted it as disrespect, but it was really because I didn't want to honour someone who would humiliate him that way. As I grew older, I recognized that however much dad lived his own life in defiance of his parents, the early conditioning he had received could never be erased. He would always hanker for their approval and endure the accompanying abuse. So gradually I went along with him without objection because I realized that my rebellion only made my father more unhappy. I pleased him by acceding, as tangible proof that he had done something right. In a way, everything I was able to accomplish became a way of adding to my father's status with his folks.

When my family moved to London, we were still impoverished. Dad's brothers pitched in to help us get settled. They helped us buy a lot near the edge of town and then contributed their time and equipment to building a house. When I arrived, the family was living in a house which had a roof but the outer shell was only framed and wrapped in tar paper. Inside, the room partitions were temporarily made of cardboard tacked onto the studs. The floor was still the original subfloor. As we accumulated more money, we gradually fixed up our house. But for the first couple of years in London, our house was little more than a shack. It was often a source of embarrassment for me.

The first car our family bought was an old Model A Ford. I'd be delighted to have one to drive today, but then it was considered a piece of junk compared to the snazzy new models everyone else was driving. Dad decided that the best way to get our garden in shape was to add mulch in the form of leaves every fall, so he built a box for transporting leaves which was mounted on the rear bumper of our Model A. This made us look even more conspicuous as we drove through the streets after work at night looking for leaves to shovel up from the curb. As a self-conscious teenager, I often hunched down in my seat, so that no one would recognize me.

HIGH SCHOOL DAYS

At Leamington High School, I had thrown myself into school activities. Even as a first-year student, I felt at home and comfortable. But at London Central Collegiate Institute in the heart of the city, I was an outsider, a country hick who had just arrived. Most of the social groups in my class had already been formed the previous year, so the only people I knew were my cousins, and they too were like strangers to me.

The summer I arrived in London, my cousins tried to introduce me to their friends. Both Dan and Art were outstanding athletes, and I certainly was not. My father had always regarded sports as frivolous. In his mind, if I wasn't studying, then I should be doing chores, not playing games. My cousins came by one day and talked me into playing football. I had never even seen a football game and had no idea what to do. After a few plays, no one even bothered with me. But I continued to run down the field, and at one point I ended up wide open. The quarterback threw the ball to me and it bounced right off my chest and dropped to the ground. Everyone was disgusted! My football career was temporarily finished.

London Central Collegiate was the oldest high school in the city, and the northern part of its constituency represented old, established money. Central was a very cliquish school, dominated by a chic in-crowd. It was also a school of whites. There couldn't have been more than a dozen members of a visible minority in that student body of about eight hundred during my four years there.

My sense of not fitting in at Central was as much an expression of my deep-seated insecurities as it was the existing social order in high school. The people with whom I eventually hung around were, like me, at the fringes of the school's social life. Our major bond was that we had no one else to be with. The terrible thing about being an "outie" is that all you want is to be part of the in-crowd. For me, it was an awkward, unhappy time, and I never felt at home in London.

Being poor and having to work instead of hanging out after school or on weekends exacerbated the separation I felt from my peers. Both of my parents worked and I was expected to go straight home from school to do chores. On weekends I worked for Suzuki Brothers, so I didn't have time to hang around with other kids and share in the culture of youth. Dad also brought me up as a serious boy. "If I were to die now," he would tell me, "I want you to be able to take over as head of the family." Of course, he never mentioned that mom had done pretty well on her own during the war.

But worst of all, I suffered from the scourge of youth in the fifties — I was smart. My good grades became a further source of alienation. Academic ability was not a quality held in high esteem by students at Central in those days, and very few students went on to university. In contrast, although Leamington High School was a rural school, I hadn't been aware of any prejudice against students who excelled in their courses. On the contrary, my class at Leamington included several students who had also skipped a grade or two, so I was in good company.

At Central, however, school spirit was focused on sports and social events like dances and the annual variety show. The social elite of the school was dominated by football and basketball players and cheerleaders. And there were other prejudices. Anyone who sang in the glee club was rebuffed by the social elite, and if you belonged to a debating group, something I enjoyed, you were beyond redemption.

I was torn by conflicting pressures: the herd instinct to be like everyone else and my father's demand that I get good marks. Once, in class, a teacher asked each of us to recite our marks. I was horrified and only mumbled that mine had all been firsts.

The major tragedy of the atmosphere at Central was that it didn't nurture one's sense of pride in academic excellence or push one's abilities to the limits. There was no joy in learning. Classes lacked challenge because teachers were too harried to worry about students like me, who were doing well on their own. I found little need to do homework because the teachers' lectures in class gave us everything we needed for the exams. When I finally ended up in an outstanding college, I suffered greatly from my lack of discipline, poor concentration and terrible study habits. In hindsight, the greatest loss in those high school years was missing out on the excitement of discovery of new ideas and the rewards that come from achievement after a lot of hard work and challenge.

High school was simply too easy for me and one result was that there were huge holes in my education. I had hated history because I couldn't see the point of memorizing endless names, dates and points in various treaties. I didn't read for pleasure and was never challenged to. Although my father read voraciously, his primary fare was fishing magazines. Perhaps I also suffered because there had been so little to read in Slocan during the war. It was only when I was in graduate school that I discovered the joys of reading.

High school is an unsettling time if for no other reason than puberty. Testosterone was surging through my system and affecting not only my body, but my brain. I spent most of my time lusting after and daydreaming about girls. Relationships with my age group took on an entirely new dimension. And adults, especially my parents, were no longer the benevolent protectors, but often seemed like terrible prison wardens.

For me, the normal social awkwardness of puberty was compounded by my sensitivity over my racial difference. The only time I ever dared to ask a white girl out during high school was when I was in grade eleven. She was popular and pretty, and had always been friendly to me. So I took a chance and was thrilled when she agreed to a date several weeks away. I spent many happy hours fantasizing about the possibilities. About three days before the big day, I called her to confirm everything. She broke the date! I don't remember what the reason was or what I said before hanging up, but I never summoned up the nerve to ask a white girl out again until I was in college. Years later, I happened to meet that same girl, by now a mother of several teenagers, but just as lovely as she had been in high school. When I told her how crushed I had been when she broke that date, she couldn't even remember that I'd asked her out!

The psychological costs of the evacuation and incarceration became clear during my teen years. Canada had made it obvious that, although I felt Canadian to the core, I wasn't because I didn't look like one. When I approached groups of strangers, in my mind I assumed that they were all assessing me as a "Jap." What the war years had done was to imprint me with that awareness of my difference.

In his searing book, *Soul on Ice,* Eldridge Cleaver recounts his epiphany when a white jailer ordered him to take down the pin-up of a blonde white woman. In that instant, Cleaver recognized that he had been carrying around the images of beauty and desirability imposed by the surrounding majority culture which led, in turn, to self-hatred. In a similar way, I, like all other Canadians, had been assaulted by propaganda that linked slant eyes and yellow skin with treachery, deceit and cruelty. And so the image looking back at me in the mirror became hateful, a tangible reminder of the enemy.

I didn't want to walk down the streets with my parents or my sisters because they embarrassed me by their Oriental appearance. And my *eyes* — they were a terrible source of shame. I read in *Life* magazine that Asians had developed an operation to enlarge eyes and I yearned to have this done. I wanted to dye my hair brown and to anglicize my name. Self-hate was the most terrible cost of the war years for me. Puberty is tough under the best of circumstances, but spending time as an enemy alien in one's own country can add special nuances.

THE JAPANESE-CANADIAN COMMUNITY

Before the war, my parents considered themselves Canadians and were very open to people outside their Japanese-Canadian community. After the war, disillusioned, thrown out of their home province and stripped of everything they had, they became much more insular within the Japanese-Canadian community. My parents kept in touch with every Japanese family in southern Ontario. They helped each other, shared gossip and met socially.

The postwar years were a time for most Japanese-Canadians to reestablish lives, careers and dreams. Evacuation and dispersion broke up the ghettos on the West Coast and scattered Japanese-Canadians across the country. This eventually destroyed the community, but it also opened up much broader opportunities for the younger generations.

In that postwar climate, interracial dating or marriage was unthinkable. My parents were anxious for me to meet other Japanese-Canadians of my age, but I had assiduously shunned contact with them in London. One evening my parents persuaded me to attend a small gathering. I met a few teenagers and found that in our mutual shyness there was something comforting and familiar. Our shared experiences in the camps and our social hang-ups gave us something in common. I was so anxious to have friends! How ironic that while I resented being Japanese, my fear of rejection ultimately drove me to Japanese-Canadian teenagers for companionship. I began to look forward to the occasional social event as my new Japanese-Canadian acquaintances became familiar friends whom I would call up and go to movies with. From these purely social needs, I was eventually drawn into the Japanese-Canadian community.

Japanese-Canadian kids of my age took it for granted that we had to work. Our parents all worked, and we tried to follow their example by excelling in school and taking on odd jobs. We weren't pushy or obtrusive — we just did the best we could. And it paid off. The astounding result was that Japanese-Canadians overcame the adversity of the war years and went on to become highly successful, model citizens.

The Japanese-Canadian Citizens' Association (JCCA) had been established before the war as an organization to fight for civil rights. After the war, it was a cohesive force for the community, keeping Japanese-Canadians informed and speaking out in support of claims for lost property. By the late fifties, however, the

JCCA had lost a lot of its impetus. The organization became more concerned with social events than with social justice.

But in the early fifties, the JCCA became an important part of my social life. The JCCA decided to sponsor an Ontario-wide oratorical contest in Toronto, and because of my successful speaking experience in Leamington, I decided to enter. All of the chapters in different towns held preliminary contests to select the people to go to the finals in Toronto. Three of us were chosen from London. I was the youngest at fifteen, while the other two were a couple of years older.

My dad became my taskmaster and coach. First, I went to the library and found out everything I could about my topic. Then I winnowed the information down into a speech that fit the required time. Once my written speech had passed muster with dad, I committed it to memory. Every night after dinner, dad and I would march down into the basement where I had to deliver it to him. If I stumbled or forgot a word, I'd have to go back and start all over again. He would occasionally interrupt me to alter an emphasis, change a word or phrase, suggest a gesture, and then I would have to start at the beginning again. I usually ended up in tears of frustration. But by the end, I could have been awakened at three in the morning and been able to deliver the whole thing perfectly!

To this day, I still use the same basic technique. For years after I became a faculty member, I would write out lectures and practice them out loud beforehand. For on-camera pieces, I still commit the script to memory by going over the lines again and again, just as I did for my father.

When we arrived in Toronto for the contest, the London contingent was like a group of small-town yokels. It was my first encounter with so many Japanese-Canadians since the days in the camp. In London, there were only about a hundred Japanese men, women and children, while in Toronto there were thousands.

Initially, I felt awkward, mainly because the Toronto kids were all older and seemed so sophisticated. But as soon as I got the first sentence out, the rhythm of the words, the nuances of emphasis and my body movements flowed automatically. It was a wonderful feeling that athletes or musicians must feel when they perform.

I won the contest and, to top it off, a pretty girl from London who had also been selected as a contestant came in second. It was exhilarating to beat the big city kids, but mainly I had something to take back to drop at my father's feet, thereby gaining a grunt of praise from him. But the real reward came on the way home late that night when that lovely girl put her head on my shoulder to sleep. It was the closest physical contact I had ever had with a girl and I nearly swooned in ecstasy. I figured everyone in the car could hear my heart pounding. I pined over her for months but I was simply too young for her.

I entered the oratorical contest again the next year, this time with my sister, Aiko, as one of the London entrants. She had decided to try public speaking and did it without the intensive coaching by our father. In spite of this, she came in second and this time I fell to third. It was a great achievement for her.

In the third year, I entered again, while my sister didn't. I won again and decided to quit while I was on top. I never entered another oratorical contest, but they had been invaluable experiences for my later careers in teaching and broadcasting. Winning was not so much an affirmation of how good I was, as a vindication of my belief that I could be as good as anyone else. That constant drive to do well had never given me a sense of superiority, only temporary relief from the need to prove myself. I found great satisfaction, almost pleasure, from the hard work of preparation and practice. The high point is always in the actual execution and the enormous surge of pleasure when it's over.

Through my involvement with the JCCA, especially in the oratorical contests, I became an active member. Eventually I helped organize a junior JCCA in London, Canada's first youth wing of the parent group. We set high goals of citizenship and responsibility, but really it was our way to meet members of the opposite sex.

In all of London, there were only thirteen teenaged Japanese-

Canadian girls, and two of them were my sisters. There were seven guys, a skewed sex ratio that made it pretty tough for the girls at parties. So both sexes had pretty slim pickings from which to choose. But we were a tight club that for a few years gave us the social outlet we needed so desperately. Our younger siblings were not nearly as inhibited and scarred by the war and readily integrated into the social fabric of their schools.

In our need for more boys, we sought out Joe Soga, a Japanese-Canadian who had been completely assimilated into the white community. At first he was aloof but gradually he became one of the gang. Eventually he became my best friend and my sister Aiko's boyfriend. Joe also worked for Suzuki Brothers and so outside of school, he and I spent a lot of time together. I had a good chum for the first time.

At the New Year's Eve party held by the JCCA in 1951, I realized for the first time how beautiful one of our gang was. Just under five feet tall, Joane had the classic face of a Japanese doll. Though she was not an extrovert, she was a terrific dancer and that night I ended up dancing with her the entire evening.

By the time the party was over, I was wildly infatuated with Joane. Now all of my daydreams had a real focus. Within a few weeks, we were "going steady" and doing all of the things that teenagers still do — spending long hours on the telephone, mooning over milkshakes, lots of necking, movies and picnics. At last I had a girlfriend, someone to show off with and to. Joane was a top student in her high school and was also a good artist, pianist and basketball player. She had a wide range of interests compared to me. She patiently struggled to teach me how to jive, waltz and tango. From grade eleven to the end of high school we were inseparable, and after I finished college we were married.

As I've said, my father had very specific requirements for my potential mate: she *had* to be Japanese-Canadian. But even dad had to be realistic about the chances of finding one among eleven in London, so he conceded that if I couldn't find a Japanese girl, then a Chinese was acceptable. After all, our races were related. But there were only about three Chinese families in all of London at that time, so if I could find a native Indian girl, that would be all right. And so it went — after an Indian or an Eskimo, he would accept a black girl. The first acceptable kind of white was a Jew because dad believed she would know what it was to be the target of discrimination. The absolute rock bottom, unacceptable prospect was British. Many of the people administering the evacuation had English accents, thus leading people like dad to believe it was they who were responsible. It is profoundly disappointing when we ourselves, the victims of bigotry, manifest the same intolerance as our oppressors.

OTHER PASTIMES

During the summers and on weekends, I worked for Suzuki Brothers building houses. I started on the concrete gang. Back in the fifties, we didn't use ready mix. Instead, we made our own concrete by shovelling gravel into a cement mixer and adding water and cement. I was in tremendous physical shape because of my youth and the work that I did. Two of us would shovel gravel into a hopper, adding a bag of cement for each load. Others would then empty it into wheelbarrows. We poured foundations, sidewalks and basement floors. Two of us would shovel nine to twelve truckloads of gravel in a day to do a foundation. And usually we'd finish well before five o'clock!

I eventually graduated to framing and loved the job. A two-man crew would start with a hole in the ground and we would frame and pour the footings and foundation, then work our way right up to the roof in less than two weeks. It was a satisfying experience, and I still love driving around London to look at the subdivisions we built. To this day, one of my great joys is to don my carpenter's apron and do a bit of construction.

Those summers were happy times for me. I took great pride in the quality and speed of my work. If I was tarring a foundation (one of the dirtiest, hottest jobs in summer), then by god I went at it with everything I had. When nailing down a subfloor, shovelling gravel or carrying planks, it was always a race against my own standards.

The men I worked with were full-time carpenters, none of whom had gone to university. They were not impressed with academic credentials, only how well one did the job. As a Suzuki (remember, it was *Suzuki* Brothers Construction), I had to overcome their prejudices about nepotism. And I did, by working hard. I was much younger than most of the men, so we didn't socialize outside work. But at lunch and coffee breaks, they educated me with their raunchy stories about drinking, motorcycles, cars and women.

Even though it was clear I was going to go the academic route rather than remain a construction worker all my life, those men weren't envious. They judged me by what I did with them. It was the shared experiences with those men and my earlier years as a farm labourer in Leamington that began to shape my political outlook and eventual commitment to the New Democratic Party.

Although we lived right in the city, there were still opportunities to be found for fishing. The Thames River flows right through London, and I spent a lot of time there fishing for smallmouth bass, one of the most thrilling fish to hook. I got to know every pool for miles. I would walk along the river, first looking for newly moulted crayfish or "softshells" which I would use for bait. The fish found them irresistible. There were also small ponds and creeks that we would hear about, then drive out to.

Fishing for me remains the equivalent of meditation. I can go on for hours, walking and casting without getting a strike, yet have a wonderful time. While fishing, I can concentrate completely on the search, scanning the water for likely hiding places and using all of the skills I gained from my own experience and years of watching dad. It's an escape to a more intimate relationship with the environment.

I should say that we never fished just for the *sport* of it, although I can't deny that I enjoyed "playing" the fish. But they were a major source of our meat protein. I have only once fished in a catch-and-release river, one famous for its brown trout in the Catskills of New York. I found it very unpleasant to reel in fish whose mouths were pitted and scarred from previous encounters with fishermen. I couldn't see the point in cranking them in just to let them go again. For me, if they aren't to be eaten, there's no reason to catch them.

The Thames River was unbelievably rich. One of my favourite spots was below Springbank Dam. It was a long bicycle trip but well worth it. I would always catch something even though I was never sure what I might pull in. I have taken calico and silver bass, carp, pickerel, and channel catfish from the river. Our family relished channel catfish — they were huge, often exceeding seven or eight pounds. They are easy to skin and clean and have few bones. The meat is dark and oily, and mom would marinate them in sugar and soy sauce and broil them. Catfish *teriyaki* is unbeatable.

In the years since my youth, industrial pollution, sewage and farm run-off have radically changed the Thames River. In only three decades, those rich populations of fish have disappeared, and the coarse fish that have managed to survive are so laced with chemicals that they are poisonous. It seems to me that nature is screaming with warnings about our impact on the environment. We think wistfully of what has been lost and dismiss it as "the price of progress." It's about time we started redefining progress.

In that Model A car, dad and I drove all over southern Ontario. Freed from the limited range of a bicycle, we explored up past Orangeville, Grand Bend and Nottawasaga. Once we were fishing in a lake in the Bruce Peninsula and dad had dropped me on shore while he stayed in the boat. I was pushing through heavy brush when suddenly I heard a sound I had never heard before. I knew instantly what it was — a rattlesnake! I froze on the spot and immediately began to quake with fear. I looked down and there, a step away, was a rattler about two feet long. I have a complete blank on what happened next, but dad claims he heard all kinds of yelling and banging. He rowed to shore in time to meet me zooming out of the brush.

When I calmed down, I realized I had lost my fishing gear and so we cautiously retraced my path. We found a dead snake with its head bashed in and my fishing rod and reel a good ten feet from where I had stopped. I must have flung the rod away in a reflex action once I'd spotted the snake. Then I'd beaten it to death in a panic.

I deeply regret what I did because as dad says, a rattlesnake is a gentleman who warns you of his presence. They are being exterminated by human encroachment. That particular snake was minding its own business when I blundered onto it, and even gave me plenty of notice to keep out of its way. Instead, I clobbered it in a fright reaction. And that's how they're being done in.

We discovered a lovely pond a few acres in size behind the Veteran's Hospital in south London. Today, there's a housing development around it, but back then there was no road access. So we would drive in as far as we could and then lug our home-made wooden punt the rest of the way. The edge of the pond was marshy and reedy, so it was impossible to fish from shore. There were lots of largemouth bass, and it was there that I caught the biggest bass I've ever taken.

One summer, I was out on the lake with my cousin. As always, when nature called, I just urinated overboard. I didn't know that the lake was surrounded by poison oak to which I am very sensitive. A day later, my face, hands and genitals were covered in terrible eruptions and runny sores. My father was convinced I had a venereal disease. In those days, about all there was for treatment was calamine lotion, so for almost two weeks I was covered in a pink crust and could hardly move for fear of being driven into a frenzy of itching. But it didn't keep me from going back to the pond. I would take prophylactic measures by covering myself with ointment beforehand. By and large, it worked. But I never did urinate out of that boat again.

By the time I had reached grade thirteen, my self-esteem had grown and I had matured in my confidence. Going steady with Joane had lifted my frustration and sense of belonging. At school, I had a few friends but none of us was part of the in-crowd. One had come to Central in grade eleven; another was an awkward boy of Slavic origin. We weren't classic "nerds" or a group of brains; we were simply an aggregate of guys who didn't have a social place in the school. We didn't spend much time together out of school, but in school we hung out. At the beginning of grade thirteen, my friends decided that I should run for president of the student body. My initial inclination was to refuse the chance because I figured there was no way I could win. I didn't want the humiliation of losing. Dad remonstrated, saying that unless I tried, I would never go anywhere: "There's no shame in trying your best and not making it. There will always be people better than you."

So I decided to try for it. There were only three people on our campaign team, but that was enough to make posters, paint slogans ("You'll Rave about Dave") and decorate our Model A. It was all great fun. The campaign ended with speeches in front of the entire student body. Here I was at a definite advantage because of my oratorical experience. Bob Taylor, my campaign manager, first went on stage in drag and made a hilarious and lascivious introduction which put the students in a good mood. Implicit in my speech was the understanding that I was not part of the "beautiful people" who set the social scene for the school. I was more serious. I appealed to all the "outies" to vote for me. Since all of the other candidates were "innies," they split the vote. We steamrollered the opposition and I won the position of president. It was a good lesson in people power. While the in-crowd were the envy of everyone else, there were far more of us outsiders.

As I neared the end of high school, my relationship with dad began to change as I became a more self-motivated, confident individual. Though dad had treated me as an adult since our Leamington days, he had always had a profound influence on my interests and my standards. His clearly defined expectations of me led me into many of my boyhood achievements. Whether it was attaining high grades, excelling at public speaking or being a top-notch fisherman, it was my father whom I needed to impress.

For years, I loved to wrestle with my dad on the living-room floor. At one point when I was in high school, I flipped him over and pinned him down fairly easily. Just as a dominant stag eventually has to make way for a stronger, younger male, dad knew I was now physically bigger and stronger than he was. He never challenged me to wrestle again.

In a sense, those innocent tussles were a symbolic test of my approaching manhood. I was rapidly nearing a new stage in my maturation. I would soon be making my own decisions and taking responsibility for my own life. And with this, my emphasis would shift from fantasies about girls to the training and disciplining of my mind.

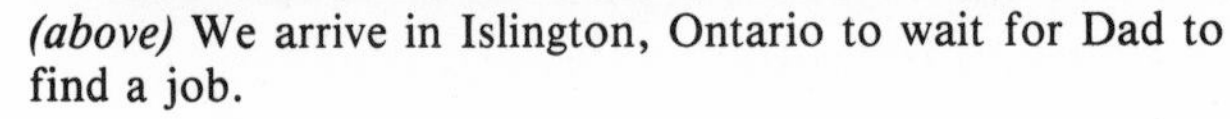

(above) We arrive in Islington, Ontario to wait for Dad to find a job.

(top right) These are harmless wampus snakes common in Leamington.

(right) This was one of my boxes of insects. Dad made the display cases for me and this won a prize in a hobby show.

(below left) This live snapping turtle was later released.

(below right) Sport II in Leamington, the second in a series of dogs I called Sport.

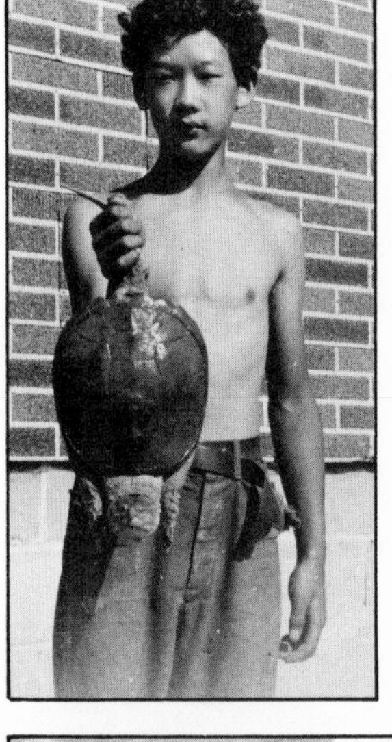

Trophies won in JCCA Oratorical contests.

By now, all but Dawn are teenagers. We're outside our home in London.

September 1953. The family car was all decked out for my campaign to become president of my high school.

A carp caught in the Thames River.

CHAPTER FOUR

AMHERST COLLEGE AND THE CLASS OF '58

THERE WAS NEVER ANY QUESTION that I would go on to university. My parents had faith that if I worked hard and did well, the opportunity would arise. They still carried the tremendous respect for education that the Japanese have. The Japanese word *sensei,* which means "teacher," implies much more — it is used with great respect and honour. To my parents, education was the magic key to changing one's status from an outsider to a respected Canadian.

The attitude of my parents was unusual at the time. In the fifties, university was still not an option for the vast majority of students. Many quit at or before completing grade twelve because of the tough government exams in grade thirteen. Of those who had the desire to go on to university, many took two years to get all of their pass marks in grade thirteen.

Of those of us who completed grade thirteen, it was assumed we would all go on to Ontario universities like the University of Western Ontario, Queen's, or the University of Toronto. Some students who couldn't pass grade thirteen went to universities in the United States. So it was our prejudice that American universities were mediocre. We didn't know there were differences between a Harvard University and a state teacher's college. We didn't understand the enormous variation in quality of American schools, in contrast to Canadian schools which are far more uniform. And that was my state of ignorance when I applied to Amherst College.

During the Christmas holidays of my last year in high school, I ran into John Thompson, who was the son of the Dean of Business Administration at the University of Western Ontario. John and I had been in the same class at London Central Collegiate. I always thought of John as an "innie" with brains. He got away with being smart because he was also very handsome and a star athlete. At the end of grade twelve, John left Canada to attend a small liberal arts school in Massachusetts called Amherst College. When I met him in London that Christmas, he was on holiday after his first semester. When I asked him how it was, he responded enthusiastically that "Amherst is the perfect place for you." He urged me to apply.

Amherst College is one of a large group of small undergraduate liberal arts schools dotted across the United States — Swarthmore, Grinnell, Reed, etc. — which are renowned for their academic excellence. It was said that unless one scored in the top 5 percent on the College Board Entrance Exams, one needn't apply to Amherst. Only one of every fifteen applicants was accepted.

I didn't know anything about Amherst at the time, but I had enormous respect for John. If he said it was a good school, I was willing to give it a shot. I thought I would have a huge advantage over American applicants by having an extra year of high school. I applied and took the College Board Entrance Exams. I was accepted and awarded a huge scholarship of $1,000 a year. (This was when you could buy a new car for $600 and live on $2,000 a year.) I have to admit, at the time I received the scholarship, I didn't know how special Amherst was and more or less took my acceptance for granted. With my parents' blessings, I nonchalantly decided to take the Amherst opportunity. Mostly I saw it as an adventure. I had no inkling that it was the most important decision I would make in my academic career.

The admissions office's policy was to accept students from broad cultural, geographic and socio-economic backgrounds, and who demonstrated a wide range of talents and interests. After I found out what an elite school Amherst was, I always believed the only reason I got in was because I was Asian and Canadian. The admissions office also tried to weigh ability in activities like acting or sculpting on a par with athletics to make up a diverse and interesting group of people. At that time, one in every five students was on a large scholarship.

It is a measure of the enormous changes of the past three decades that the class of '58 profile now seems incredibly homogeneous. But in comparison to the students at London Central, my fellow freshmen seemed an incredibly diverse group. The student body came from all over the United States. I was the only Canadian and one of two Asians. There were four blacks and a high proportion of Jews. But the predominant numbers were upper-middle-class WASPs.

ORIENTATION

Amherst, about a hundred and twenty miles directly west of Boston, is a lovely New England college town nestled in undulating hills. In full fall colours and in winter snows, it is like a picture postcard. The distinctive clapboard houses with their colourful shutters are typical of the buildings in that part of the country.

Amherst College was established in 1821, and is therefore a relatively young school by New England standards. It was started as an all-male private school on the initiative of Congregational ministers and local people who literally built the school with their own hands. The college soon established a tradition of academic excellence. It began as a community effort and remains the pride of the town today.

Early in September 1954, I left home on my grand adventure. I was starting college! It was exciting and a bit daunting. I arrived for Freshman Orientation a week before the upperclassmen returned and formal classes began. All freshmen — "frosh" — were housed in three dorms, most in single rooms. As might be expected, we formed our best friends in our respective dorms, often in those first days of school.

Orientation Week was highly organized at Amherst. A few upperclassmen who lived in the frosh dorms as advisers were there to explain the facilities and to warn us what to expect. There were evening parties for us held by the president of the College and various deans. There were social events — mixers with frosh from the neighbouring girls' schools, though I was too shy to do more than gawk.

During the day, we were introduced to the campus facilities — the gym, infirmary, chapel, etc. Faculty explained the various programs and curricula. They constantly stressed that we were part of a history of high standards. The feeling of history and superiority at Amherst was overwhelming. In a sense, it was all indoctrination — but it worked! Throughout our Amherst careers, we were proud of the school and felt a strong responsibility to strive for excellence in all we did. Each of us was a top student. We had been accepted at Amherst and therefore we were special. Every student was expected to do well and there were tremendous resources available to help any student who might have problems. There were psychologists, tutors and buddies to help us through any tough moments.

By the end of Orientation Week, we were prepared for classes and the return of the upperclassmen. As lowly freshmen, we had to spend the first week of classes abjectly wearing a purple and white (Amherst's colours) beanie and obey any demands made by upperclassmen. We had to have a "proper attitude" toward upperclassmen, meaning one of total servility. We had to get out of their way when they passed. And we had to be prepared to perform a trick, make them laugh, sing the school song, carry books, or do errands when so commanded. The pranks were harmless and, in retrospect, fun. At the time, I quaked with fear of humiliation, but the experience gave my class a sense of camaraderie and shared experience.

Today, it appalls me to see students, often from small towns, arrive at Canadian universities homesick, lonely and frightened, without support facilities to help them adjust. As freshmen, they are thrown into huge lecture rooms and have to take labs and tutorials presided over by graduate students. It's an assembly line where individuals become ciphers funnelled through the system to justify university budgets. Of course, it's not valid to compare a small, highly selective private school such as Amherst with a huge state university, but I believe that the large Canadian universities could be much more supportive of their students.

As soon as I arrived on campus, I looked for a job. Amherst was very expensive even with my scholarship and I needed to work during the school year to supplement my income from summer jobs. After a few weeks, a chance came up to work the morning shift as a busboy in the dining rooms. Although students were housed in dorms and fraternity houses, we ate in two dining halls in one building. This provided a way of mixing students and minimizing small cliques. Breakfast was served from seven in the morning. We would show up a half-hour earlier to bring out the milk and butter, make toast, and so on. We wore little busboy jackets and were assigned sections in the dining room. It was our job to be sure that the tables in our areas were cleared of trays and wiped down, salt and pepper shakers kept full, and sugar and napkins put out. It was a snap. There were long periods when there was no work so we could snooze or do homework. But we would work hard during peak times. Usually we'd be finished long before a 9:30 class, but we were always paid for three hours. At seventy-five cents an hour, I was able to earn enough to cover most of my living expenses beyond room and board for the entire four years.

There were no drugs back then. Marijuana and LSD were simply unheard of. The fifties were a time for liquor and pranks. The two best pranks in our dorm were the smearing of a toilet seat with gobs of peanut butter and pouring dissolved clear gelatin into a toilet bowl where it hardened invisibly.

Academically, Amherst was a real challenge for me. I, who had hardly had to crack a book and breezed through grade thirteen, expected to coast through college on the basis of my extra year of high school and what I thought was a superior school system in Canada. In fact, I found myself surrounded by students who were not only much brighter than I, they were far better prepared and motivated. They were the cream of the American crop. Almost a quarter of my class of three hundred students had been valedictorians of their schools!

My first day of classes was like being at a horse race when the starting gates are opened. All of the other horses bolted ahead, while I was still waiting to figure out what was going on. They tore into their assignments like children after treats. My high school habits haunted me. I sat through classes unable to concentrate. I was still daydreaming about girls. I didn't know how to use the library, had no discipline to read assignments and didn't know how to organize an essay. As the recipient of a scholarship, I had to remain in the top 20 percent of the class or I would lose the money. I knew right away that education at Amherst was not going to be a breeze.

The key to academic excellence is good teaching and small class size. From my very first courses, class sizes were never larger than twelve to fifteen students. Freshmen did have lectures in large lecture halls in some courses, but the tutorials were always smaller and taught by professors, not teaching assistants. Good teaching was expected, not as a token goal, but as the basis on which faculty promotion and tenure were granted.

A sense of camaraderie between students and faculty was stressed. The courses were at once intimate and terribly demanding and brought out the best in us. Each course had an allotment in its budget for "entertainment" of the students. In the upper classes, most professors would use it to buy beer for evening sessions. For our last class in a botany course, our professor took us on a field trip to "watch the buds open." He led us on a long hike through a woods, occasionally exclaiming, "See, there's a bud there! Oh, there's another one over here!" We were mystified until we reached a cabin where he had stashed a huge cache of ice-cold Budweiser beer.

THE AMHERST RENAISSANCE MAN

An eclectic undergraduate education was Amherst's mission. Amherst College only gave Bachelor of Arts degrees. It was assumed that Amherst graduates were not trained for a job, but were honing their minds on a broad range of subjects. We were trained to be well-rounded "Renaissance Men," destined to become leaders in whatever profession we eventually selected. But first we were to learn more about the world around us. So, although I knew I would concentrate in biology, I still had to take many courses in the arts and humanities. In fact, during my freshman and sophomore years, only 40 percent of my courses were in science, while in my junior and senior years, science courses could not exceed 60 percent, even though I was in honours biology.

Almost all the information from my college science courses is hopelessly out-of-date, inaccurate and irrelevant. Yet the fact is, I still use skills I learned then. That's because the emphasis was on process, not nickel-knowledge "facts." We were taught to ask questions — how do we know this, what is the evidence, how can we test this, etc. Now, more than ever, students need those skills. But with so much scientific research going on, most undergraduate courses are designed to impart the most current information.

My humanities courses opened new horizons for me. Bits and pieces of those humanities courses have proved useful time and time again in my later career. I still treasure courses I took in political science, psychology and world history. One of the most worthwhile subjects I studied was classical music. In my milieu in London, kids thought the greatest kind of music was country and western or pop. I really believed that people who said they like classical music (we scornfully called it "long-hair" music) were putting on airs. Taking that course was a revelation. Through this music appreciation course, I encountered Palestrina, Bach, Handel, Beethoven, Mozart and others. We had to memorize pieces and know them well enough to identify a short snippet played during an exam. I spent long hours in the listening room, blissed out on some of the greatest music ever written.

At Amherst, we were expected to strive for academic excellence. Making the Dean's List for scholastic achievement brought with it privileges, like cutting classes, leaving early for holidays, and social status within the college. Jocks still scored a lot of points in social standing, but one received just as much respect if one were in theatre, played in a school orchestra or wrote poetry. Competition to be chosen as a member of one of the small singing groups was fierce and there was immense prestige attached to making the school glee club. Those who played chess, belonged to language clubs, joined debating teams or worked on the school paper earned respect and recognition for these activities. It was an inversion of my high school years.

Amherst took the well-rounded "Renaissance Man" image quite seriously. It was considered important to be able to speak well in public. All Amherst men were required to take a two-semester course in public speaking in their sophomore year. Each of us gave three short speeches per semester and were critiqued by the teacher as well as our fellow students. It was regarded as a bit of a lark since no one ever flunked, but I took it very seriously as a chance to improve my speaking ability. The bonus of that course came at the end of the year. The best speakers in each section gave their talks at a local prep school for girls, where it was reputed that rich, beautiful girls would give up their virginity instantly to an Amherst man. We went panting in anticipation, only to spend an incredibly boring evening chit-chatting over canapés and tea after our performances.

We were expected to be accomplished in matters not only academic, but cultural and physical as well. No one would receive a degree without being able to swim two lengths of the pool. I can remember a black student from St. Louis, a star basketball player, who struggled to complete that swimming requirement in his very last semester.

All students had to take physical education classes or play on a school team for two years. Away from my father's influence, for the first time I was actually able to try my hand at sports. I found out that not only did I enjoy sports, I was even reasonably well coordinated. In fact, I swam competitively for the freshman team. John Thompson, my good friend and mentor, was a star swimmer and had urged me to try out. On my first day, the coach pushed us into windsprints using a kickboard. I refused to quit until I ended up puking my guts out in the gutters. I stuck with swimming, but hated the practices and suffered from nervousness for days before a meet.

I also tried out for lacrosse and played on the freshman team. It was a real thrill for me to suit up and play competitively. I kept thinking, "Me — a jock! Think of the girls I could have had in high school." Amherst had recently discovered rugby as a way of cheating on a ban on spring training for football players. It was still a new sport on campus so I went out and made the team. Through physical education classes, I learned to play tennis and volleyball.

But my greatest love was touch football. I played it right on through graduate school. Even after I became a faculty member at the University of British Columbia, I played for years with the students. My great hero was Jimmy Brown, the magnificent fullback who carried the number 32 for the Cleveland Browns. Long after I grew out of daydreaming about girls, I still spent many happy hours imagining myself as a smaller version of Brown.

SOCIAL LIFE

When I first arrived at Amherst, I wanted desperately to fit in with my schoolmates. But, as usual, I felt completely out of my element. While Amherst made a serious attempt to recruit scholars from low socio-economic families, the fact is that what makes most elite schools work is money.

Today, the full cost of tuition, room and board runs to $14,363 a year at Amherst. This prohibitive sum is offset by scholarships, the money coming from alumni donations and corporate gifts.

The college's endowment fund as of April 1986 stood at a mind-boggling $250,000,000, one of the highest per capita in the United States. Those funds are an enormous resource that can be dipped into for special projects — a new building, a special chair for a prestigious scholar, merit increases for faculty. The endowment becomes a cushion to the school for unexpected or sudden needs. It is also used to fund the many scholarships for undergraduates.

Sons of alumni were given special consideration for admission. And often, when it was obvious a student was goofing off and sliding through (we said they were getting a "Gentleman's C"), you could usually guess that his father was an alumnus and a heavy donor.

If I had gone on to an Ontario school, I would not have been out of place at university, but here in an expensive and elite American college, most of the students came from upper-middle-class backgrounds of generations of university graduates. And I was the first member in my entire family to go to university. For the first time in my life, I felt self-conscious of my unfashionable clothes. My schoolmates seemed to take wealth and privilege for granted. Even ordinary conversation emphasized the gulf that separated us. Many of my classmates had gone to private high schools. They had travelled extensively throughout North America and abroad, and went to plays and symphonies. They drove their own cars and brought fancy hi-fi sets. And they were well read. Of course, everyone wasn't like that, but to me it seemed that way.

At this point in my life, I finally began to transcend my feelings of always being an outsider. For the first time rather than being cowed and intimidated, I learned to appreciate the differences between myself and my classmates. I saw my exposure to my fellow classmates as a tremendous challenge and opportunity, a chance to learn, and gradually my self-doubts began to subside. I also bought the ideal of the Amherst man. I believed the school was special and that by being accepted, I was good. While it was immediately obvious I was not going to walk into class and be a star student, I was not afraid of hard work and

I figured I could hold my own. My father had always urged me to reach beyond myself. Though he was tough and demanding, his love and support served to reinforce my sense that I was worthwhile.

I always had some inner belief in myself that gave me the temerity to feel that I could be as good as anyone else, regardless of the differences in our backgrounds. My grandparents had come from the poorest of classes and had been subjected to terrible discrimination, but they managed to maintain their unself-conscious sense of pride and confidence and passed it on to my parents and then to me. Even in my most unhappy times as a child in Slocan and an awkward teenager at Central High, the core of my self-esteem was always nurtured and kept intact by my family.

John Thompson eased my adjustment to Amherst. While he had a large group of friends and an active social life, I could always seek him out if I needed advice or help. But this didn't happen often; I tried to make my own way. When I discovered a Japanese-American name on the same floor of my freshman dorm, I sought him out. What a surprise! He was a huge man, over six feet tall and built like a linebacker. Honolulu was his home, and he came from a privileged background, his father being a wealthy doctor. He had gone to Punahou, an exclusive private school in Hawaii, and was just as aware of the socio-economic gulf that separated us as I was. He tolerated me but never hesitated to point out the cost and source of his clothes and fancy possessions. But I tagged after him like a puppy. Years later, when I went to Honolulu for a job interview at the University of Hawaii, I dropped in on him. It had been almost ten years since we had last seen each other but he couldn't have cared less about my visit. I finally realized that what I had thought had been a friendship was, in fact, of my own creating.

Amherst College was an all-male school, but its location was particularly strategic. Aside from the University of Massachusetts, which was on the other side of town, the women's colleges, Smith and Mount Holyoke, were only seven and ten miles away respectively, giving a sex ratio in the area of at least three women to one Amherst man. We got to know those roads intimately.

The upperclassmen at Amherst were, to us, like eagles who would hover above the hubbub of the first weeks of school and then swoop down to pluck the best girls from the freshmen classes at Smith and Mount Holyoke. They seemed so suave and sophisticated in comparison to us freshmen. Each school prints a book of the freshman class which features each student's picture and a short description of vital statistics. The freshmen books from Smith and Mount Holyoke were as valuable as gold. There was at least one on each floor of the dorm. It was on the basis of pictures in those books that lots of blind dates were set up. As the weeks went by, guys would write comments beside the pictures. Girls marked "hot to go" or "easy mark" must have wondered at their popularity. The first thing I did was to look through the book for female Oriental faces. In the fifties, there weren't many attending Ivy League colleges, perhaps one or two in each class at Smith or Holyoke. So things weren't better for me in college than they had been in London.

Most of us had come to Amherst with steady girlfriends back home, but there was far too much pressure to date to remain faithful. Most of us had "understandings" with our girls that we would date while at Amherst. During the first months I kept sending a steady stream of passionate letters to Joane. She had moved to Toronto to study chemistry at Ryerson, but I was confident she would stay with me. I don't know how many high school romances survived the four years at Amherst, but I'd bet there was a direct relationship with the proximity of the girlfriend to Amherst. Guys with girlfriends in neighbouring towns could bring them to Amherst for special events or whip over to see them on long weekends and keep the fires burning. What the guys did in the meanwhile was kept top secret.

It was a time of double standards. Today, in the aftermath of the turbulent sexual revolution of the sixties, it's difficult even for me to reconstruct the sexual morés of the forties and fifties.

Those were conservative times, and we males actually expected to marry a woman who was still a virgin, although we were also supposed to be sexually experienced through our own hell-raising.

In that first year, I had a half-dozen dates I can remember. Three were Chinese girls. I took John Thompson's sister Anne to the Spring Prom, but I wouldn't dare try anything fresh with her. It was like taking my own sister.

My first date with a white girl was an arranged blind date and it was a disaster. I was assured that "she's *not* ugly and she's *very nice.*" I knew she wouldn't be a beauty, but I also knew damned well that I was no great shakes either. I found the immediate evaluation based on physical appearance distasteful, which is not surprising considering my self-consciousness about being Asian and not being fashionably dressed. But when we went out as a group, I was acutely conscious that my schoolmates had immediately judged her as a "dog" and written her off. It turned out that she *was* a very nice girl, but under these circumstances it didn't matter. We were both tense and awkward and glad when the evening was over. I never went out on a blind date again.

The entire social structure at Amherst was built around fraternities or "houses" as we called them. Well over 95 percent of the students belonged to fraternities and lived in them during their junior and senior years. Each of the thirteen fraternities had its own distinctive flavour — Alpha Delts were blond and good-looking, Dekes were goof-offs, Thetas were jocks, etc.

The fraternity houses were the centre of student activity from sophomore year on, and the college depended on them for their social function as well as housing. It would cost the school a bundle to offer comparable facilities, so the fraternities were in a powerful position to maintain their status. The admission of large numbers of women has changed all that, but in the fifties Amherst's fraternities dominated the campus.

Fraternities are a strange anachronism on a campus priding itself for its tolerance and liberalism. Amherst's great rationalization and boast was that its fraternities had no racial or religious restrictions and that there was "100 percent rushing." All those who wanted to join a fraternity were guaranteed they would get into one. This seemed to indicate an open system, but it wasn't. Fraternities are, by definition, discriminatory and exclusive. It is true Amherst fraternities were not allowed to have discriminatory clauses and that a handful had actually broken away from their national organizations by admitting blacks, Jews and Asians. The fact is that most frats retained the blackball — the power of one anonymous person to deny admission of a prospective candidate — and there were houses which would never admit an Asian, Jew or black. But there were so few Asians and blacks in each class that some fraternities did not have to use the blackball. Most fraternities were all white.

John Thompson belonged to Phi Alpha Psi (called Phi Psi), a fraternity that had been kicked out of the national organization of Phi Kappa Psi in the forties for taking a black student. Phi Psi had always had black members. Through John, I had met many of his fraternity brothers early in my freshman year and I liked them. I was delighted when they asked me to join. As it turned out, ten of the fraternities offered me membership, and I was blackballed in the other three.

The most repugnant part of fraternities occurred in the spring at the end of the "rush period" when fraternities chose their members. Almost everyone would be selected by a fraternity by the end of the week. But a handful of people, called "turkeys" in the argot of the time, would have received no offers. Everyone knew who had joined which fraternity, and was equally aware of who hadn't. Since Amherst guaranteed 100 percent rushing, every person had to be placed somewhere. So, representatives of each fraternity would meet and they would bargain over taking the last people. Quotas were lifted and one fraternity might offer to take a certain "turkey" if the others would divide the rest.

It was as if these last people were no longer individuals. From then on these people bore the indelible stigma of the judgement that they were losers. To this day, I can name the last five people in my class to join. They carried this judgement around with them for the rest of their campus careers. Amherst's proud boast of

100 percent rushing slid over the pernicious effect of selecting those last people.

I hated this process, and in later years when I was a more confident upperclassman, I urged our fraternity to do a nonselective rush, to let people choose *us.* We could take them on a first-come, first-accept basis until our quota was filled. I felt that the very fact that we had all been admitted to Amherst made each of us worth getting to know. But this was too radical a challenge to the notion of selective exclusivity. So in my senior year, I quit the fraternity and moved back into a dorm.

For two years, though, I was an active member of Phi Psi and made solid friendships. We had great parties at the houses on weekends when a keg of beer would be tapped. It was at Phi Psi that I learned to love touch football. There was always someone ready to toss a frisbee, although there were no commercial models then so we used metal pie plates or tops of potato chip cans. Phi Psi was academically oriented. We usually had the top grade average of all houses. And Phi Psi *always* won the competition for the Fraternity Sing.

It was in Phi Psi that I first heard a black use the word "nigger" to describe himself. He used it with irony, showing he had said it deliberately and trusted us enough to use it. He knew how sensitive I was about the word "Jap," and when I would come into the house, he'd announce, "Hey, guys, here's the slant-eyed Jap!" He taught me that it's important to judge *who* is using a word and *how* it is used. The word "Japanese" can be said in a way that is every bit as insulting as "Jap." That doesn't make the derogatory word acceptable, but used among friends, it's a token of trust. That black fraternity brother also taught me not to take myself too seriously. He mocked my most sensitive insecurities and helped me face them with a bit more perspective. He taught me that unless we can pull back and laugh, we can be consumed by bitterness.

I learned a further lesson from another black fraternity brother. In my third year, I moved into Phi Psi where I shared a room with three others, one of whom was black. It was a difficult year, and at times there was a great deal of tension between us. At the end of the year, I realized that I just didn't like the black student and he no doubt felt exactly the same about me. But I had been so determined to do the right thing that I had never given him a chance. I only saw him as *black,* so I was determined to have him as a roommate and to like him. That was the only way I saw him, as a black person who, by being my roommate, was a symbol of *my* liberalism. I never gave him a chance as an individual. If I had done that, I would never have roomed with him. It was only in admitting that I didn't like him that he became a person to me.

During my junior year, Alfred Kazin, the renowned literary critic and author who taught at Amherst at the time, spent an evening talking to us at Phi Psi. At one point he told us that he read several books a night. Most books he wouldn't bother reading entirely. He claimed that within the first twenty pages or so, he knew whether it was going to be any good. Then he challenged us by saying that we should be reading a book a day. I was floored by what I felt was an impossible goal. I felt bogged down by the enormous amount of work and couldn't see where I'd have time to read for pleasure. But I never forgot the goal he set.

BACK HOME

As a Canadian, I was classed as a foreign student. So during my first Thanksgiving at college, I took the opportunity to be billeted out with an American family during the holidays. It was a remarkable experience. Americans are genuinely hospitable, warm and open. I was welcomed as an adult and my ideas were listened to with respect. During the Thanksgiving dinner, which was a large rollicking affair, members of the family talked and argued enthusiastically about world affairs. My head snapped back and forth as if I were at a tennis match.

It was a performance, like none I had ever seen. There was the *mother* of the family holding forth about politics just as loudly as her husband. Not only did she express her opinions unequivocally, but she dared to contradict what the men were saying. This was far different from our Suzuki family gatherings in my grandparents' farmhouse. In *my* family, women quietly did the work and talked among themselves. They never raised their voices, never engaged in a political discussion and never contradicted their mates in front of others. That Thanksgiving experience opened my eyes to a whole set of new attitudes toward family life.

Each year I returned home for Christmas and Easter breaks, and for the summers. I talked endlessly about what I was doing at Amherst to my parents, who listened eagerly. Even with my scholarship and work as a busboy and in construction, college put a terrific economic strain on the family. Mom and dad couldn't afford to visit Amherst until my graduation, but they knew it had been the right choice for me. In typical Oriental fashion, they and I have always felt a great debt to John Thompson and his parents for making it possible.

Like all parents, mom and dad were clearly very proud of me and revelled in my growth. In most Asian families, children carry a heavy burden of parental expectations. I certainly knew that I was expected to improve my lot in life over that of my parents, but it wasn't an oppressive weight — it was simply the way of life.

My parents may have been unusual Asians in that they never stressed the need for economic security, to make a lot of money. Like my father, I've never been able to manage money — our wives have done that for us. There was much more emphasis on fulfillment — work should be enjoyed and there should be pride and satisfaction in working hard and doing a job well.

During my second year at Amherst, I had a violent argument with dad that marked a major change in our relationship. In my last year in high school, my father had scolded me for something I had said, not because of *what* I had said, but because he felt other people would not like it. Later, when he found out that people agreed with me, he told me how proud he was of me. It shocked me that he had put *appearances* before *principle.* It was a minor incident but it mattered greatly to me because my great idol had been tarnished. Unthinkingly, dad had let his concern over public opinion affect his attitude. He had compromised a principle that he drummed into me all my life, of standing up for what one believes.

Two years later, when I came home from Amherst, I articulated my resentment over this incident in an angry exchange with my father. In the past, dad had always overpowered me in our arguments — if not with his superior reasoning, then by his superior position. He demanded to have the last word. This time it was different. I now had two years of college under my belt and was more self-confident and aggressive.

When I attacked him vigorously, to my horror he didn't respond in kind, but slumped in his chair, covered his face and wept. It was the first time he had ever cried in front of me and it signalled yet another change in our relationship. He had conceded he was wrong by humiliating himself in front of me. "Men never cry," he'd always taught me. I felt no joy in that victory, but it freed me to love my father in a different way.

Whenever I returned to London, I resumed going out with Joane and we spent all of our free time together. She too returned to London for the holidays. During my sophomore year, I received a disturbing letter; Joane had met a guy in Toronto and was getting seriously involved with him. I too had been dating at college, but I didn't expect Joane to change *her* allegiance. Her letter hit like a bomb. All of my competitive feelings were aroused. To add to the complications, I was also taking Physics 22, a course that was regarded as the one that separated men from boys. It was required for everyone who was a pre-med student. Many a potential medical career foundered on Physics 22. I was having a hell of a time with it and was in danger of failing. Joane's letter came on top of that.

So I decided to quit school. I called my dad, told him there was a good chance I would flunk a course and thereby lose my scholarship. He urged me to reconsider. He told me that even if I lost the scholarship, he would do everything he could to finance my stay at Amherst. He suggested I could make up the course in summer school, a prospect that horrified me with the loss of face and enormous costs involved. He then said, "Look. There's no shame in failing, so long as you can look me in the eye and tell me you've done your best." That crafty man! He knew I couldn't do that.

But hormones and immaturity had their way. I took a train to Toronto to rekindle the flames of my romance with Joane. It was a humiliating experience. She was in the flush of a new romance, and I failed to make an impression. A few days later, a slightly wiser and vastly more humble student crept back to Amherst where I begged to be reinstated. I was put on probation for the rest of the year.

That summer I returned to London to work for Suzuki Brothers. Joane had a job in Toronto so I spent a lot of time hitchhiking there from London on weekends. Through heroic efforts, I was able to persuade Joane that I was the preferable suitor and I suddenly found myself "pinned." At Amherst, when a girl was given someone's fraternity pin, it was a statement of matrimonial intent, just short of engagement.

Like a feudal lord, I had successfully reestablished hegemony over my "territory." But once the flush of regaining Joane's commitment wore off, I wondered whether I had done the right thing. I'm sure Joane had as many misgivings, but the thrust in the fifties was towards getting married and "settling down."

FINDING DIRECTION

Fortunately, I scraped through that sophomore physics course and retained the grade point average I needed to keep my scholarship. Those first two years had been terribly humbling. I had learned that I was far from the smartest guy around and, that among scholars, hard work and discipline mattered as much as intelligence. Slowly, I learned how to apply myself. By my third year, I was ready to make decisions.

There were two biology courses which were inspirational to me. One was Embryology taught by Oscar Schotté, who had studied under the Nobel Prize winner, Ross G. Harrison. Schotté, was a character, a short, fat man with a heavy accent. He was also an irrepressibly enthusiastic teacher. He spent enormous amounts of time teaching us the pure mechanics involved in transforming a nondescript egg cell into a multicellular creature made of numerous organs and tissues. How does an egg do it? It's a central question that has preoccupied scientists for centuries.

The way classical embryologists study the process of early development is to do experiments on frog embryos at different stages, pickle them and cut cross-sections for viewing under a microscope. By scanning embryos from successive stages, scientists try to map out where and how cells and tissues move and change. It's tedious, painstaking work, but Schotté made it come alive. He kindled my interest in the phenomenon of early biological development.

The other course was genetics taught by Bill Hexter. Hexter was a marvellous lecturer whose style and approach reflected the training he had received at Berkeley under Curt Stern, one of the giants in genetics. When I began taking the course, I knew nothing about the subject. Hexter taught it like a detective story; he would describe an observation in heredity, then pose the question, "What mechanism would explain this phenomenon?" He would then suggest theories or models that would, in turn, lead to ways of testing the ideas. Finally, he would show us some actual experimental results that led to a definitive conclusion. We gradually gained a well-grounded understanding of inheritance.

For the first time in my life, I sat in a class completely enthralled, my mouth hanging open in astonishment at the beauty of the insights and the elegance of mathematical precision absent from most other areas of biology. I couldn't wait to get to the next class. Gone were the days of daydreaming. I was hooked and asked Hexter if I could do an honours thesis under his direction. Now remember, I had not distinguished myself in my first two years at college. I was in the top 20 percent, but barely. Trying for an honours degree meant doing a research project, writing a thesis and lots of hard work. I felt lucky and proud when Hexter chose me as one of his three honours students.

In the fall of 1957, my senior year of college, an epidemic of Asian flu hit campus. Classes were decimated. I, too, finally came down with it and had to make my way to the college infirmary. While I was there, the announcement that the Soviet Union had launched Sputnik came over the radio. It was electrifying! People had actually put an object into space.

Just as Japan's economic prowess in the eighties has shocked the United States, the Russian launch of Sputnik sent a bolt of fear through American society in the fifties. The Russians were scoring political victories in a turbulent Africa and in Southeast Asia, but their space spectacular woke us up to their progress in science. Suddenly there was a spate of articles about the high social status of scientists in Russia, the rigid selection of students and their intense training in scientific disciplines. And just as Americans are comparing themselves to Japanese today, back then everyone tried to learn from the Russian example.

That first space success marked the beginning of a period of intense activity as Americans faced up to a formidable Russian competitor. Their lead seemed insurmountable. Out of that period, NASA — the National Aeronautical and Space Agency — was born, and Congress began to pour billions of dollars into science and education in order to catch up.

The space program was clearly based on ideological competition. But it had a ripple effect that stimulated all of the sciences, including biology. For budding scientists like me, it was a golden period of opportunity. Big scholarships were established and university science faculties began an explosive period of growth. Our future was one of total optimism. We didn't have to worry about *whether* there would be jobs, only *which* one to look for. There were more jobs than people to fill them. We took it for granted that doing science was an intrinsically good thing, not because it fulfilled some national goal or fed into the economy. The search for knowledge was a new frontier.

It had been my intention to become a doctor after graduation. I applied to and was accepted by a medical school in Canada.

But genetics changed all of that. I was assigned to investigate a theory that Hexter had developed to explain a rare genetic event in the fruitfly, *Drosophila melanogaster*. It was a clever experiment and when I arrived in the fall, Hexter had all of the stocks of flies ready. It was just a matter of setting up and carrying out the experiments to get the data. That suited me fine and I learned plenty of genetics as a "fly pusher."

There were about a dozen of us in honours biology. Each of us was required to deliver a lecture to faculty and students each semester. During the first semester, we had to choose a published scientific paper and report on it to the group. In the second semester, we had to discuss our own thesis projects. Ours was a group with diverse interests, ranging from ecology to genetics to biochemistry. The challenge was to present the material in a way that would be understandable but still exciting to everyone. I found that I had a knack for doing this and while the group didn't normally applaud at the end of talks, they gave me a big hand after my first presentation. It was exhilarating! I had taken a very esoteric paper in fruitfly genetics and explained it in a way that most people not only grasped but could recognize for its elegance.

I didn't realize it then, but that was the first hint of where my ability lay. All those years of explaining to dad what I had done at school had an effect. In reading a difficult paper, I would try to think of how I would explain it to my parents so they could see why it was exciting or important. It's very simple really, but scientists tend to get caught up in fine details and try to qualify everything to be accurate. Too often, the central question and the passion and excitement are drained away.

During the seventies, I hosted a program on "The Nature of Things" about fruitflies and my research. One of my colleagues, a fruitfly geneticist himself, came up after it was broadcast and exclaimed, "Thanks a lot. For the first time, my kids got excited about fruitflies!" I thought, my god, if you can't make your life's work exciting for your own children, it must be pretty tough to be an inspiring teacher.

I tend towards hero worship, and the authors of the scientific papers I read filled me with awe. I knew I could never aspire to their heights but I thought that if I could get a Ph.D., my niche would be to teach genetics in a university. As a teacher, I would be able to enjoy the excitement of genetics vicariously. I turned down the medical school offer and applied to graduate school. My father tells me that my mother wept in disappointment that I gave up medicine in order to study fruitflies!

In that last year, my grades shot up so that my marks were among the top in my senior year. The Biology Department gave me a very high grade and recommended that I be granted an honours degree *magna cum laude* (with great honours). Because of my mediocre beginning, my grades weren't good enough. One needed at least an 86 percent average for all four years to be granted a *magna.* The three students in my class who graduated *summa cum laude* (with greatest honours) had a four-year average of over 90 percent. I graduated *cum laude* (with honours) which for me was achievement enough, as it required above 80 percent for all four years.

Amherst College had provided an environment in which academic excellence was a valued goal, not a social detriment. For the first time in my life, I had been immersed with students who challenged me to excel in reaching for new ideas. My horizons were extended in all directions by the demands of a liberal arts tradition. Unlike science students today who must digest an enormous amount of specialized material, I was afforded the luxury of lifting my sights beyond the immediate confines of scientific disciplines.

I was impressed with the quality and tradition of an Amherst education. In turn, it instilled in me a sense of my ability — I gained confidence that I could compete and contribute. It seemed only natural to go on into science. I felt a love for genetics that burned within me, and I owe much to my esteemed teacher, Bill Hexter. He had inspired me, and become a treasured friend, and had encouraged me to go on.

When I graduated in June 1958, I was a radically different person from the one who had arrived just four years earlier. The upheaval in my schooling caused by dislocations in British Columbia and my ennui in London had been compensated at college. Now I was a young man prepared to go out into the world and find out how good I really was.

(above left) This is the London "gang" dressed to kill. My cousins Art and Dan Suzuki are at the left, then Howie Cagawa, Vic Uchiyama, my best friend, Joe Soga and me.

(above right) A carpenter in London. Rev. and Mrs. Nakayama are author Joy Kogawa's parents.

(left) Joe Soga and me with channel catfish caught in the Thames River.

(below left) Joe Soga and I were in fine shape from construction.

(below right) This is my picture in the freshmen book at Amherst College.

(above left) Fraternity brothers who dropped in to London. From the left, that's John Zinner, me, Harold Haizlip and Kit Schemm. Note how "cool" we were in 1955.

(above middle) Dad and I hit the jackpot in trout. Note the crewcut that he has worn ever since Pearl Harbor.

(above right) Summer '58. Fish biologist for the Dept. of Lands and Forests in Sudbury. Fred Possmeyer (now a professor at the University of Western Ontario) was my partner.

(below) 1958 — A college graduate.

CHAPTER FIVE

STARTING OUT IN SCIENCE

THE BEST PART OF SCIENCE has always been in the *search.* The fun is in experimenting, designing new tricks and often getting totally unexpected results that then lead in entirely new directions. The longlasting rewards come from the total involvement the work demands, the excitement of discovery and the camaraderie of those who share in the search.

There is a false notion that scientists set out objectives, and then in a linear fashion go from point A to B to C to a cure for cancer or some other objective. That's how a bureaucrat would like science to work, but if that were the case, most of life's problems would have been solved long ago. Science simply doesn't happen that way. It involves going where your curiosity leads you, taking you into blind alleys and down side streets, often landing you far away from where you intended.

People, including scientists, don't agree on the meaning of good science. My belief is that scientific ideas are tentative and it's not so important that they're right as that they suggest further experiments. Once a scientist has an hypothesis to explain certain observations or data, this immediately suggests experimental tests or directions for further observations that will prove or, equally important, disprove the hypothesis. If the data negate the idea, then a new concept has to be invented. Confirmation of the hypothesis, on the other hand, leads to further refinements and experiments. This is how scientific "truth" is established. Thus, a concept like evolution or the double helix of DNA can lead to a radical revision, reassessment and progression of experiments.

Sometimes we can follow a promising idea that seems to explain everything. Then years later, it can collapse when more information is obtained. Science is really in the business of disproving its current models or changing them to conform to new information. In essence, we are constantly proving our latest ideas are wrong.

Governments put enormous pressure on scientists to do work that has some kind of practical spin-off. Increasingly, scientists feel constrained to couch their proposals and aim their work towards that end. But that is not how science is actually done. Good scientists, asking good questions, will find interesting answers, attract resourceful young people, and in the end have ideas that lead to practical spin-offs, although the final application may be totally unanticipated.

People often ask me what I have actually contributed to science. Like most scientists, my contribution has been an incremental addition to the vast body of scientific literature. But the work for which I have gained some recognition has demonstrated that it is possible to probe development and differentiation by using a certain class of mutations. They are hereditary defects which are perfectly normal at one temperature but completely defective at another. By blasting such mutants with a temperature shock at different stages in development, the thermal effects can be examined. This process has become a very useful tool in many areas of genetics.

However, to fully understand our lab's contribution, we have to have a perspective that encompasses the long historical roots of genetics. The exciting work going on in the field today is just the latest in a long history of the study of heredity.

THE ROOTS OF GENETICS

Genetics is said to have begun as a science in 1900, when fundamental laws of heredity that had first been reported in 1867 were rediscovered. But people had been interested in questions about inheritance all the way back to our earliest records of human thought. Having children, after all, has always been the central fact of our biology.

Because human beings lived in small bands of nomadic hunter-gatherers for 99 percent of our history, the health of each newborn has always been critical to the prospective survival of every group. So since prehistoric times, people prayed and practiced elaborate rituals to try to ensure that healthy babies were born. Defective infants were probably allowed to die by either passive neglect or active infanticide. Since many of these defects had a heredity basis, these actions were a crude form of genetic selection.

In fact, genetic selection takes place long before the time of birth. It is estimated that for every live birth as many as three fertilized eggs will have failed to implant or have been aborted spontaneously. This kind of selection has worked very well. Any parent knows how incredibly elaborate the process of reproduction is. And through rigid selection, this process has been programmed into our bodies to take place all by itself.

Hereditary differences, such as skin colour or facial features, have always been important in human affairs. They can form biological marks that distinguish enemies from friends. Within a group, blood relationship is of crucial importance in defining kinship patterns. Inheritance of social status, occupation and property are often biologically based. All of this tends to make heredity and biological relationships of consuming importance in every society. Even today, in highly traditional societies like India, matchmakers remain important and prospective mates are carefully selected according to elaborate rules. Much of this practice may be socio-economic in intent, but there is also the underlying importance of procreating children who are strong and healthy.

Long ago, people were thinking about heredity and how it might be influenced. Some of our earliest records of writing are in cuneiform tablets found in the Middle East. Dating back some six thousand years, fragments exist that prescribe dietary rules and ritual activities to be followed during pregnancy to ensure the birth of a healthy baby. Obviously, the current interest in using amniocentesis and ultrasonography is just the most recent practice in the long history of trying to monitor and determine the newborn's well-being.

Of course, the indisputable evidence that early people understood and used principles of heredity is that they domesticated plants and animals. Ten to fifteen thousand years ago, pockets of people in many parts of the world discovered the value of cultivating plants and taming wild animals. Within the remarkably short period of a few thousand years, virtually all of the plants we grow and animals we breed today were domesticated. And as those early farmers and ranchers became more sophisticated, they began to select for characteristics that were useful — size, appearance, heavier fur, docility.

The most insightful principle to emerge from early agriculture was "like begets like," a concept so important that we find it written in many books of wisdom like the Bible. Even today, plant and animal breeders use this simple principle to select and breed desirable types. There were more sophisticated insights as well. Ancient Jews recognized the heredity of haemophilia, a disease which we now know is caused by a gene on the X chromosome. They observed correctly that the defect is passed only from the mother, but is expressed exclusively in sons. Thus, Jewish scholars decreed that when a woman gave birth to two sons who died of bleeding after circumcision, all subsequent sons were exempted from the sacred ritual.

In the days of ancient Greece, Lycurgus founded the city-state of Sparta and dictated that human heredity should not be left to chance coupling. So he set up a panel to pass judgement on which men and women were fit to have children. Before a newborn baby was a month old, it had to be brought before the committee to be evaluated. Those infants deemed unfit were left to die at the foot of the volcano, Mount Taegetus. Plato suggested that Athenians should be bred with the same care as hunting dogs and birds.

So human beings have long used hereditary principles in animal and plant breeding, while wondering how to apply those practices to their children. But it has only been within the last century that we have really understood the mechanics of heredity and acquired the tools to do something about it. This study of heredity as a science actually began in the last century.

The Augustinian monk, Gregor Mendel, raised garden peas. By breeding them carefully over several generations, he was able to derive principles that apply to all plants and animals. Unfortunately, no one realized their significance when he published his results in 1867. It was only at the turn of the century, long after his death, that Mendel got the recognition he deserved. Probably, as so often happens, his insights were premature in the sense that there wasn't a sufficient body of knowledge to "make sense" of his conclusions. By 1900, the biological community was ready and genetics burst on the scene as a brand new science that enjoyed spectacular success.

The key insight made by Mendel was that we don't inherit actual traits such as hair colour, skin colour, eye shape, etc. Instead, the *potential* for those traits is transmitted from parents to children. We pass on a set of instructions called *genes* which specify the development of the traits. Every gene is present as a pair, one of each pair coming from each parent. So parents equally contribute genes to their children (with some exceptions such as the genes on the X and Y chromosomes). A simple characteristic, such as blood type, may be controlled by a single pair of genes. Other traits, such as skin or eye colour, are produced by the interaction of many different genes. And for something as complex as a brain, or more abstractly intelligence, the number of genes involved in shaping it is so great that all that can be done is a statistical analysis of the characteristic's inheritance in a population. At the level of individuals, there are no obvious patterns for the inheritance of intelligence. Of course, there are severe defects such as Tay Sach's disease and cretinism that result from mutations in one or two genes. But for the vast bulk of people falling under a broad range of intellectual ability, there is no simple hereditary pattern.

We now realize that each generation represents a new combination of genes inherited from its parents. For simply inherited characteristics, it is relatively easy to determine the basis. Mendel's insights reveal a pattern of heredity that allows us to predict the frequencies of different genetic combinations in offspring. But it's a prediction still based on probability, not absolute certainty.

It's one thing to talk about abstract entities — genes — as if they are symbols in an algebraic equation, but do genes actually correspond to a physical entity in a cell? Only two years into the twentieth century, scientists realized that the statistically predictable behaviour of Mendel's "factors of inheritance" correspond precisely with the visible movement and distribution of *chromosomes* in cells. A strong circumstantial case was made for genes being carried on chromosomes, thereby giving a physical basis to units of inheritance.

The first years of this century were a remarkable period in biology. A flurry of reports quickly showed that Mendel's laws were universal, and where exceptions seemed to be found, they in turn led to new insights. The young field of genetics was astonishingly successful. It was in this climate of excitement that Columbia University professor, Thomas Hunt Morgan, shifted fields. He was interested in embryology, a field in which one tries to understand how an egg transforms itself into a complex, multicellular organism made up of many different cells, tissues and organs. It's a scientific area dealing with the mechanism of development and differentiation. Morgan knew that development in every species follows a pattern characteristic of that species. So development must come down to a matter of heredity. Morgan became the father of the line of geneticists to which I became a member.

THE FRUITFLY: A PERFECT GENETIC ORGANISM

Many embryologists who study early development choose amphibians — frogs and salamanders — as their favourite organisms because they are easy to rear and produce large eggs in great numbers that can be readily observed and manipulated.

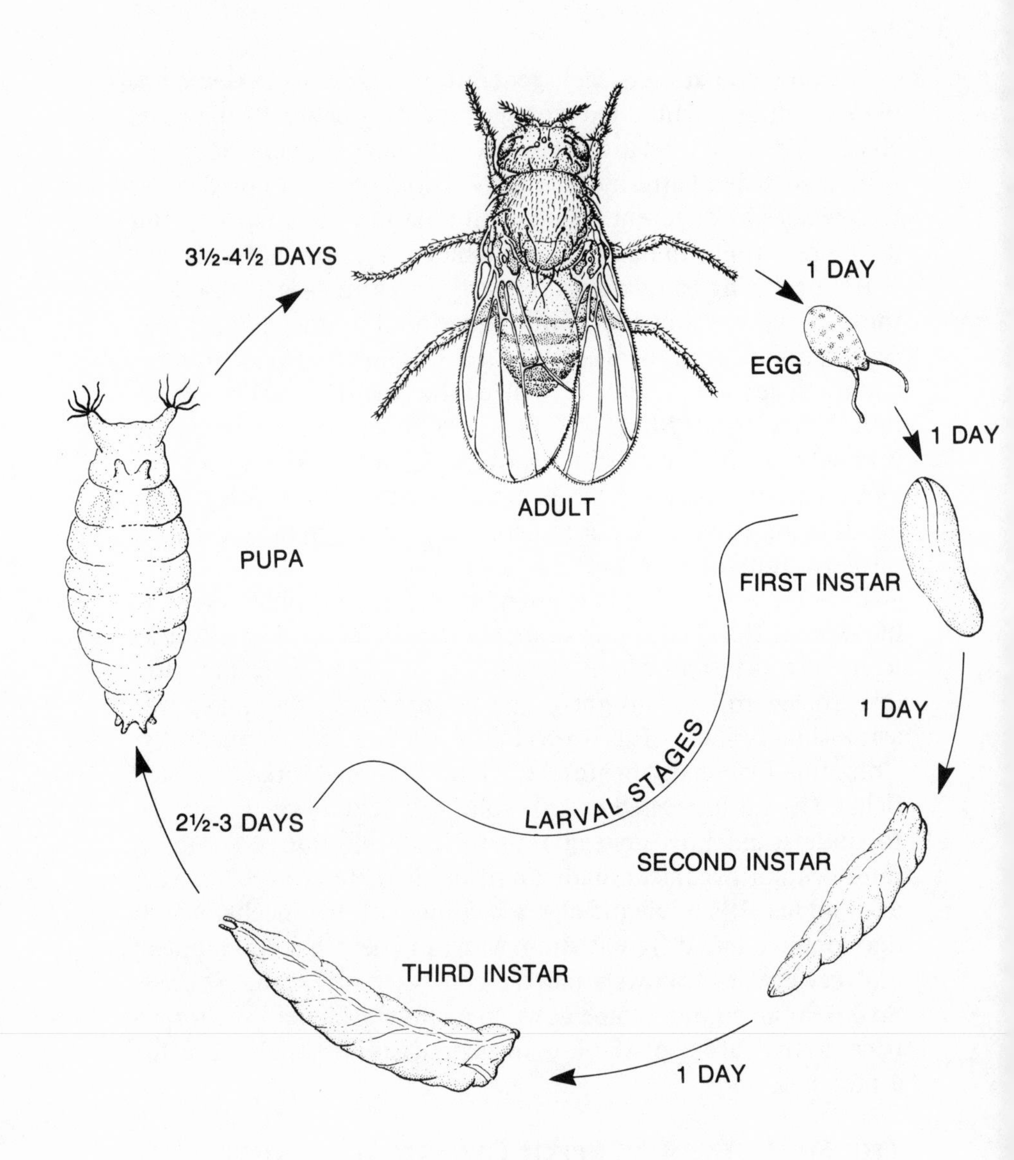

FIGURE 1

The life cycle of the fruitfly
at a temperature of 25° C.

What do geneticists choose as an ideal organism to study? They need an organism that reproduces large numbers of offspring because genetics is largely a statistical science. This rules out mice, frogs or plants, since they would take a lot of effort, space, equipment and money. So the ideal creature should be small, prolific and easy to raise.

It is said that a colleague of Morgan's at Columbia University was annoyed by fruitflies hovering around his dessert and suggested in a pique that Morgan should study them. They were the common fruitfly, *Drosophila melanogaster* (which literally means "dew-loving, black gut"). We are familiar with these insects as clouds of tiny flies that seem to appear out of nowhere near bowls of ripe fruit. In fact, fruitflies feed on yeast, not the fruit itself. Morgan did select this tiny, unprepossessing animal and it turned out to be a prescient choice. Many years later, it was discovered that flies also possess giant chromosomes, an unexpected bonus that even Morgan could not have anticipated.

What's so interesting about a fruitfly? A fruitfly, like many insects, undergoes a dramatic metamorphosis during its life. (Figure 1.) After hatching from an egg, the worm or larva is a perfectly good and functional organism with a nervous and muscular system. It moves and responds to smell, light, gravity and temperature. It eats and defecates. In short, it does everything much the way we do, short of reproducing. After a week or so, when it has shed its skin twice and grown large enough, the larva pupates and undergoes an incredible metamorphosis. When it emerges from the pupa case, it is a totally different creature with eyes, wings, legs and reproductive organs. An intelligent being from outer space would be hard-pressed to identify the larva and adult as different phases of the same species.

The way the larva sets the stage to "give birth" to the adult form is remarkable. When the larva hatches from the egg, it already carries small clusters of cells called "discs" which are "programmed" to form the adult structures. (Figure 2.) In the larva, they merely grow in size as an amorphous mass.

How do we know they are already programmed? If a disc is

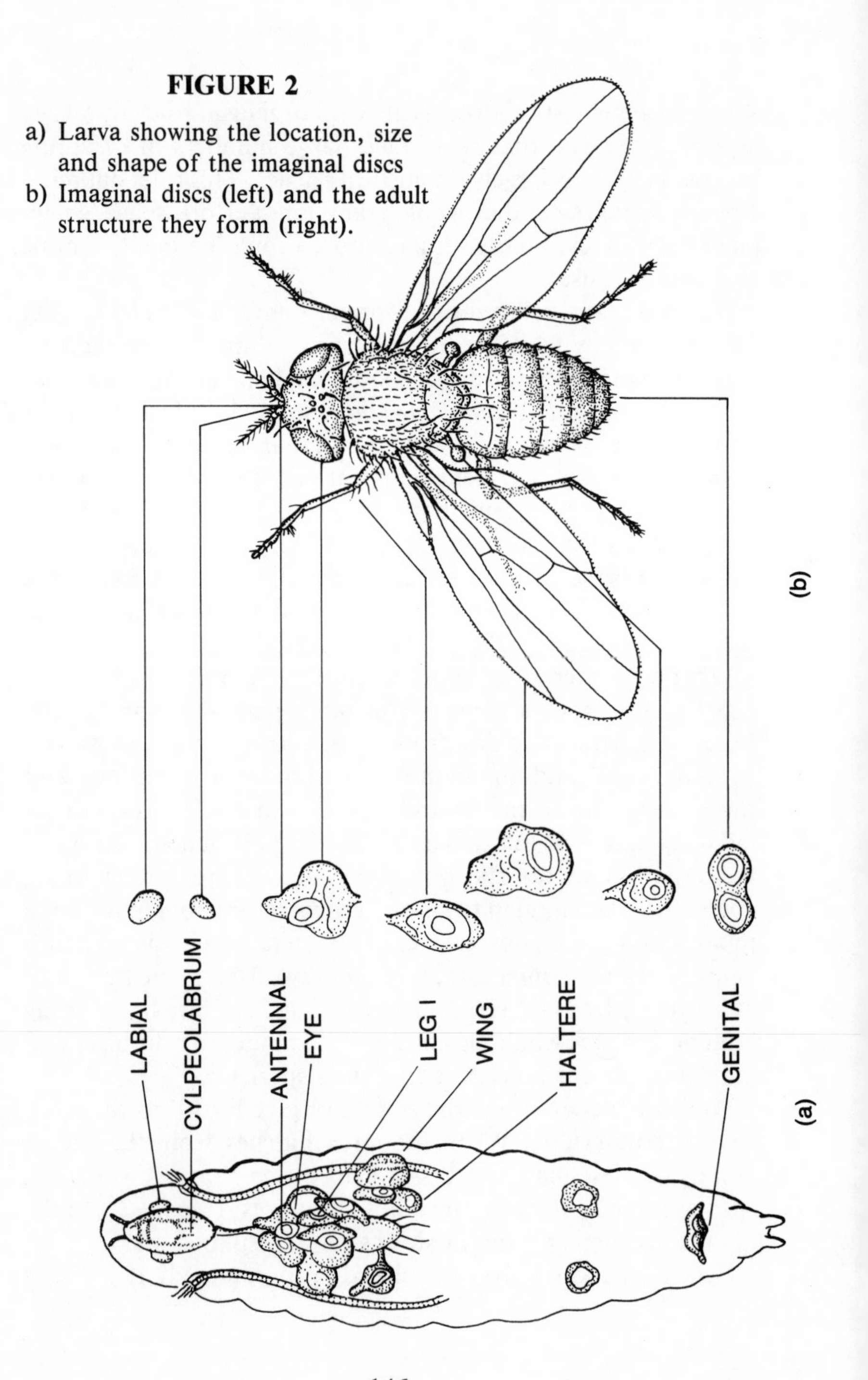

FIGURE 2

a) Larva showing the location, size and shape of the imaginal discs
b) Imaginal discs (left) and the adult structure they form (right).

removed from a larva and injected into the gut of an older larva which then pupates, the injected disc will form the structure for which it was programmed. Hence, if an eye disc is injected into a larva's gut, then when that larva becomes an adult, it will carry an eye in its abdomen. Of course, the transplanted eye won't function because it's not hooked up properly.

When a larva pupates, a hormone is released that acts as a trigger to activate the discs to become the adult structure for which it was programmed. As the larval carcass breaks down inside the pupal case, each disc independently differentiates, one disc forming an eye, another a wing, a leg and so on. The leg disc is like a collapsed telescope which forms the leg when each section slides out to form one of the leg segments. The wing disc forms a bag like a collapsed balloon. When the adult emerges, the bag is pumped up with fluid which then drains out through veins. The two sides then come together to form a flat blade two cell layers thick. As the differentiating discs meet, they stitch together to form the adult.

A newly formed adult is tightly crammed within the tough pupal case. How can it get out? To help, an empty bag on top of the fly's head is inflated and pops open the top of the pupal case. Once the fly emerges, the hydraulic bag deflates, never to be used again. It's astonishing to reflect on the complexity and precision programmed into those tiny creatures. A biological miracle takes place every time a fly is born!

The challenge for geneticists is to determine how these events are genetically programmed. The nervous system of the larva seems to be the same as that of the adult, because if a larva is conditioned or taught to respond to a stimulus, the adult reacts the same way without being trained. Flies have a complex repertoire of behaviour that can be used to probe the nervous system. They must walk properly, know when to fly and land, find food and mate. They have complex organs to detect the environment — compound eyes which fracture a scene into small overlapping parts and antennae to detect chemicals and air pressure. Flies can sense gravity and temperature and exhibit a periodicity of behaviour that suggests an in-built "clock."

Mating of flies involves a complex sequence of behaviour. A male follows a female and flutters his wings to make a seductive buzz while bunting the female's abdomen. As she walks away, he follows and may move one wing out in a scissoring motion. This can go on for minutes before the male either gives up or succeeds. When the female accepts the male, she allows him to climb on her back and curl his abdomen down to lock genitalia.

The entire life cycle of a fly takes about ten days at lab temperatures. A female will mate within twelve hours after emerging from the pupal case, and during a three- or four-week adult life can lay several hundred eggs. These flies are so prolific that a small shelf of bottles can contain tens of thousands of them. The flies grow readily and cheaply in cultures containing a jelly-like mix of nutrients (mainly yeast and sugar) and agar. They are easy to handle and readily anaesthetized with ether or carbon dioxide. Once they are knocked out, they can be easily moved about with a fine paintbrush. Since every part of their bodies is controlled by genes which can be detected in different states, the fruitfly is a geneticist's dream. In over twenty years of studying them, I have never failed to find them fascinating.

THE DAWN OF MOLECULAR GENETICS

By the time I went to college in the fifties, all of the ground work of "transmission" genetics — the way genes and chromosomes are inherited — had been worked out. The glamour of *Drosophila* had faded as a new generation of microbial geneticists exploited the rapid growth and vast numbers of cells (millions in one drop of fluid) that could be handled in a single test tube. Bacteria and viruses, by virtue of their simplicity relative to flies, gave us very detailed insights into how genes work and are organized.

In 1953, Jim Watson, a young American scientist, and Francis Crick, a British physicist, published their blockbuster, a short paper proposing a molecular structure for deoxyribonucleic acid or DNA, the genetic material. It's not often that one can point to an experiment or paper and say that it was pivotal or revolutionary, but their paper was. As soon as it came out, we knew it had to be correct because it "smelled" right. Watson and

Crick's proposal of a double helix of two long chains coiled around each other was so simple and elegant one sensed that it just *had* to be true. It paved the way for the modern era of molecular biology by giving us a model that immediately allowed us to see how genes could encode information, duplicate themselves and change through mutation. With the Watson-Crick structure of DNA, genetics changed from a statistical science into a physical science, with the potential to actually manipulate the stuff of life in a predictable way.

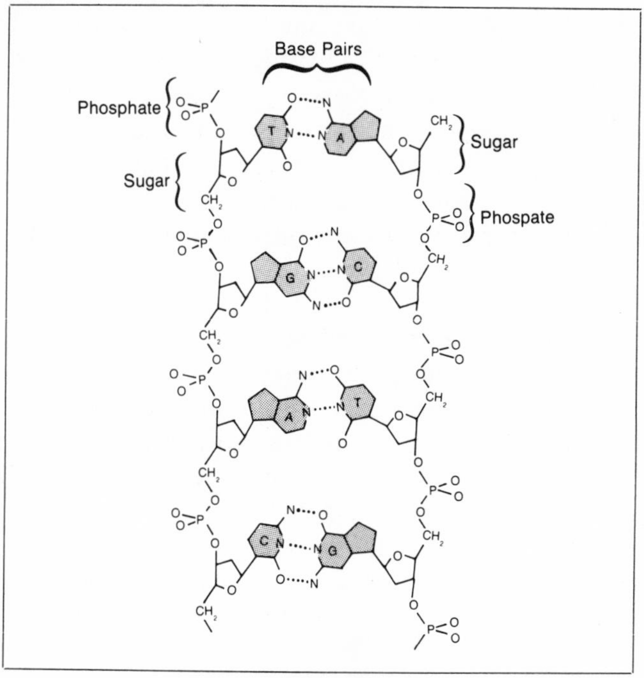

FIGURE 3

The molecular basis of DNA. Each continuous strand on the left and right is made up of alternating sugars and phosphates. From each sugar, a base is extended into the centre. Of the four different bases, A always pairs with T, and G with C. Note that there is a *polarity* in the way the sugars are arranged into two strands . When a model is made of this structure, the two strands twist around each other to form a helix.

The Watson-Crick model and its implications are simple to grasp. Basically, this model involves two chemical ropes made up of alternating units of phosphate and sugar. (Figure 3.) From each sugar, extending perpendicularly to the direction of the strands are four different structures called bases, which can be labeled simply A, T, G and C. The letters stand for the names of the bases: adenine, thymine, guanine and cytosine. These bases are defined by their chemical shapes and exist in DNA in a very specific relationship. The A on one strand is always opposite a T on the other, while the G on one side is always opposite a C. This A = T and G = C relationship reflects their pairing; they "fit" together like a hand fits a glove.

DNA is very long, perhaps millions of base pairs in length. If you look down one strand, you can see that the sequence of bases is like a message written out in a linear array. (Figure 4.) The genetic alphabet is made by those four letters: A, T, G and C. If you think that can't spell out very much, remember that we can link only two symbols — a dot and a dash — in a linear array of Morse code and spell out every word in the English language. The DNA carries information in the sequence of bases along its length. In 1961, Crick proved that the bases are "read" unidirectionally, three at a time in a triplet code. It's easy to see what a *mutation* or hereditary change is — any kind of typographical error that changes the sequence of bases.

The specificity of the base pairing A with T and G with C is the key to DNA's ability to duplicate itself. If you picture the two strands as a giant zipper, so that at a signal the two sides begin to unzip, the process leaves the two single strands with their bases hanging free. If you've eaten a nutritious meal, you've consumed the basic building blocks of DNA, so that when an A is free, a T will be inserted opposite it. On the opposite strand, the free T will be matched with an A. You can see that in this way, as the two strands unwind, they will reform a new double helix by attracting their complementary partners. (Figure 5.) The two DNAs at the end will be identical to each other and to the original

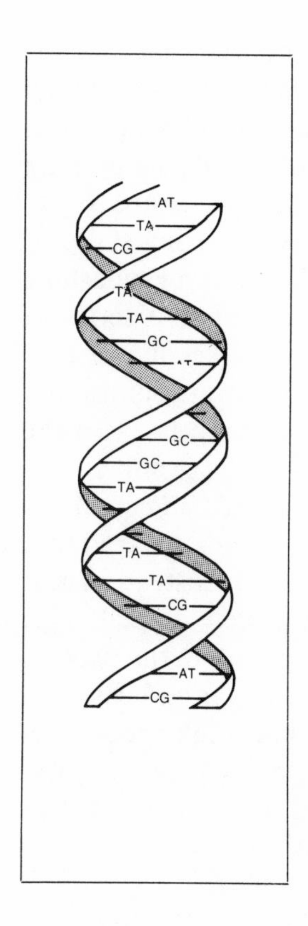

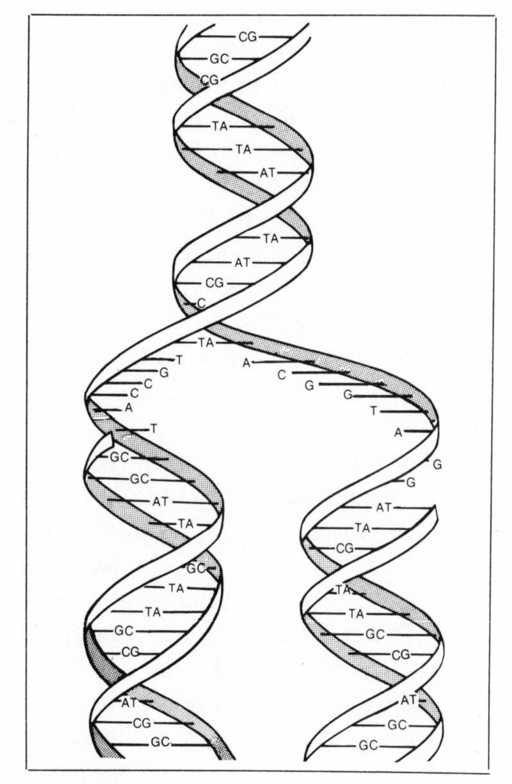

FIGURE 4

DNA molecule showing the *sequence* of bases along its length. The sugar-phosphate backbones are represented as ribbons, while the letters represent bases suspended in the centre.

FIGURE 5

Replication of DNA. The two strands of the original molecule (top) are pulled apart to give unpaired bases (centre). They attract their pairing partner and each strand forms a copy (bottom).

DNA with which we started. It's a marvellous way for a chemical substance to replicate itself faithfully.

When I started as a biology honours student at Amherst

College, DNA was known to be the genetic material and the Watson-Crick model had already been developed. But human genetics was still so primitive that the number of chromosomes in people was not yet clear. That number was in dispute because chromosomes were so small that they were therefore hard to see.

In the late fifties, T.C. Hsu in Texas was preparing slides of human cells to study their chromosomes. This involves squashing tissue on a glass slide and then dipping the slide in a series of solutions of preservatives and dyes. When he looked at one particular preparation, he was astounded to see chromosomes that seemed to be as big as sausages. Something had happened to make the chromosomes much bigger and thus easier to see and count. He realized that something "wrong" had occurred and carefully retraced what had been done to those slides. He discovered that a technician had accidentally put distilled water instead of salt solution in one of the jars in which the slides were dipped. This had the effect of causing the cells to swell up, and in the process the chromosomes stretched out and became larger and easier to see. Today, that technician's action is standard lab procedure. This is an example of what some people refer to as serendipity, but it's important to realize that the good scientist follows up and takes advantage of a chance observation.

CATCHING FIRE AT THE UNIVERSITY OF CHICAGO

I was one of three honours students in genetics in my last year at Amherst. As I mentioned earlier, the inspiration for my interest in genetics was our supervising professor, Bill Hexter. He regularly invited his honours students to his home for dinner and regaled us with scientific lore and gossip. He seemed so wise and worldly at the time, yet he was only in his early thirties. Hexter's attentiveness towards his students provided the model on which my own lab would later be built. He told us how *his* professor, Curt Stern, the great German geneticist at the University of California at Berkeley, met with his graduate students every week for a bag lunch. The idea of students being able to spend time so intimately with such a great scientist thrilled me.

Because of Hexter, Curt Stern became one of my great heroes.

In 1958, after I graduated from Amherst, I went to Montreal to attend the International Congress of Genetics which is held every five years. Stern was there and gave an inspiring lecture that I was thrilled to hear. I introduced myself to him as his scientific grandson through Hexter, and he responded with a warmth and interest that I've found is characteristic of all the great scientists.

In my senior year, Hexter assigned my project. It concerned the phenomenon of "crossing over," the physical exchange of parts between similar chromosomes during production of eggs. The experiment he designed was straightforward. It was a matter of setting up a lot of flies and counting huge numbers. I worked hard and got lots of data, but was unable to confirm Hexter's hypothesis. Still, it was good to learn the discipline of planning and carrying out experiments. By the time I had decided to pursue a Ph.D. in genetics, a semester was already over. Applications to graduate school had to be made early in the fall, and I sent mine in after Christmas.

My great ambition was to work with Curt Stern at Berkeley. Stern seemed pleased to receive my application and immediately accepted me. However, my application was too late to compete for fellowships. And without a fellowship, I couldn't afford to go to graduate school, especially as far from home as California.

Hexter then urged me to write to his friend Bill Baker at the University of Chicago. Baker not only accepted me, but offered me a job as a research assistant on his grant, providing me with enough money for graduate school.

The Department of Zoology at the University of Chicago had a distinguished history with some big names in biology as faculty members early in this century. The University of Chicago had been started by the Rockefellers who had aspirations to see it become a midwestern equivalent of Harvard or Stanford. But its location along the Midway, the site of the 1893 Exposition, was now the deteriorating junction with the black ghetto. The university had a seedy look, with none of the wide-open spaces of the Ivy League schools like Princeton. Nevertheless, it had a very high academic rating, no doubt reflecting the prestige of

Robert Maynard Hutchins, a past president, who had instigated a radical departure from traditional education. Hutchins believed in the highly motivated student who excelled at independent work. Classes were small with lots of tutorials.

The Department of Zoology of the Unversity of Chicago had coasted for a long time on the reputation of its early science stars. The building itself seemed straight out of the nineteenth century — dark and musty. One of the students kept a coatimundi, a fox-sized relative of the racoon. It was a lovely animal, gentle and shy, and it stayed out of sight in the attic during the day. At night, it would emerge and wander through the halls. One year, the eminent British biologist, Julian Huxley, was a visiting scientist in our department. As I was coming in to the lab one evening, Dr. Huxley came bolting around the corner and spotting me, exclaimed rather breathlessly, "I say, there seems to be a rather *large animal* in the halls!"

When I arrived in 1958, there were some older students who had been working on their degrees for years. They had families and other jobs. They were often aloof and kept to themselves. A few of us newcomers decided that we needed to get together and share ideas. We organized a noon seminar for graduate students and invited guests only. We'd bring bag lunches and discuss our own work or a paper, or listen to a visiting scientist. Our seminars started out with about a half-dozen students, and a lot of amused tolerance from the older people. But we found that visiting scientists were often delighted to meet informally with students and we began to have a lot of good speakers — even Nobel Prize winners. Our seminars were soon taken more seriously and, after a while, faculty members in the department asked to attend. Soon our lunchtime seminars became an important part of graduate student life.

It has always been my experience that it only takes a few people to generate an enthusiasm that becomes infectious. When I had my own lab, sometimes certain combinations of two or three enthusiastic people suddenly clicked and the whole lab was lifted up. Similarly, a couple of complainers can have a devastating effect on the morale of a group. At Chicago, a handful of us, by our own keenness, galvanized the department. Years after I graduated, I was delighted to find that the graduate student seminars were still going strong. It was a nice legacy from our motley crew.

At Amherst, I had been struck by the huge disparity in size between a DNA molecule and a chromosome. Even though we need a microscope to see chromosomes, they are tens-of-thousands of times bigger than DNA molecules. Yet chromosomes are able to come together, pair and exchange parts with a precision that goes all the way to the sequence of bases in DNA. How does the cell achieve this process? I had the temerity to think that I would like to tackle that question.

Professor Baker, on the other hand, was interested in development. He was studying a system in which the amount of pigment produced in the eye varied under different genetic conditions. My job, as his technician, was to carefully cut off the heads of flies, squash the heads onto a huge blotter-type paper and suspend the paper in a container in which one end was dipped in a chemical solvent. The solvent moved through the paper just like ink on a blotter, and when it passed the squashed heads, the eye pigments dissolved in it and were carried along. But since the pigments had different properties of size, shape and electrical charge, they travelled at different speeds and soon separated out from each other. I would then cut out each spot of pigment and measure the amount in a machine. It was tedious, but it paid my way.

Baker belonged to an elite group of fly people who call themselves "chromosome mechanics." This group delights in constructing clever ways to rearrange genes and chromosomes, tying them in rings, shifting them from one place to another. It's an esoteric skill that lets us do wonderfully clever experiments and gives a lot of intellectual satisfaction. Too often, though, it becomes a way of showing off to a small band of cognoscenti without making any important contribution to the larger world of genetics. But I found that I enjoyed this art immensely and was good at it. Amherst had taught me well. When I had arrived at Chicago, I found that in spite of a liberal arts degree, I was able to hold my own with other students.

Each week, different members of the lab would get up and report on his or her research and then we'd discuss it. It was a good way of sharing information and getting fresh perspectives on our work. During one of these sessions, a student who was finishing up his thesis reported a series of crosses. One of them involved two different kinds of X chromosomes. He observed in passing that the amount of crossing over on other chromosomes was different in the two crosses. Although he barely noted it, I was intrigued. I couldn't see why the structure of X chromosomes should affect the behaviour of other unrelated chromosomes.

So I proposed to study this "interchromosomal effect" on crossing over by constructing different kinds of double X chromosomes and studying their effects on crossing over in other chromosomes. In hindsight, I can argue that such experiments may tell us something about the spatial relationships of chromosomes during cell division. But at the time, I just knew I could follow up on a puzzling observation by applying my skills as a budding chromosome mechanic. The study became my thesis project and I looked at a whole range of aberrant chromosomes on crossing over. This didn't provide me with a definitive mechanism for the phenomenon, but each experiment led me to set up an explanation that could then be critically tested. In fact, that has been the story of my work — doing experiments, setting up hypotheses to explain the results and doing more experiments to test the hypotheses.

During graduate school, we blossoming scientists were being heavily conditioned. Our values and beliefs were being shaped around the professional ethos of the time. We were taught that science is always *good* and of benefit to humankind. To question science and its sometimes detrimental consequences (such as nuclear weapons and toxic wastes) was to be anti-intellectual and to impede *progress*. Any suggestion of proscribing research as potentially dangerous represented a fundamental attack on the scientist's absolute right to freedom of enquiry. These were "sacred" values that we absorbed, often unconsciously, during our years of graduate training.

We also absorbed lessons from what we were *not* taught. We did not learn that there may be limits to what can be "known" through science or that there might be deleterious consequences of science that should override the imperative to continue. As well, we relished the competitive, crushing criticism of others' work and believed such criticism ensured top-notch science. We heaped contempt on people in areas such as ecology that we felt were second-rate disciplines. We regarded anyone who had stopped publishing as "over the hill" and therefore no longer worthy of respect. In the enthusiasm over new ideas being proposed, it's easy to think that anyone who is no longer alive or productive couldn't possibly have been as bright as the authors of the latest papers. Such is the short historical perspective of many of science's practitioners.

Science remains an activity that is highly competitive, macho and exclusive. Its practitioners often wear blinders to questions about social responsibility, about negative effects of science and technology, about ends and means and possible limits to the scientific enterprise. People in most professions are similarly oblivious to such questions, but few areas have consequences as immense for society and its future as does science.

During my time at the University of Chicago, Doctor Baker helped to organize the "Midwest *Drosophila*" meetings. These were informal get-togethers of the small band of fly people in Indiana, Wisconsin, Illinois and Michigan. Faculty, visiting scientists and students would meet on one of the campuses for a couple of days and we would make friendships and measure ourselves against each other by presenting our work for criticism. For students it was a terrifying experience, but also very rewarding since the sessions were usually friendly and nonconfrontational.

Once we met at the University of Indiana in Bloomington where H.J. Muller was a professor. In 1927, Muller had discovered that radiation increases the incidence of mutations. Years later, this work earned him the Nobel Prize. On the occasion of our meeting, Charlotte (Lottie) Auerbach was visiting from Edinburgh, Scotland. Auerbach was another "biggie" who had opened a whole new area of genetics with her discovery in the forties that mustard gas, the potent chemical used in World War I, caused mutations. Because of her work — chemical mutagenesis — the induction of mutations by chemicals replaced radiation as the way of generating new mutations.

Lottie was formidable in scientific discussion and I was always terrified of getting into a disagreement with her. She had a rather prominent nose, and during one of the sessions she and Muller got caught up in an animated disagreement. She stuck to her guns and Muller finally leaped to his feet and exclaimed, "No, no, no. Can't you see? It's as plain as the nose on your face." He instantly realized in horror what he had said and tried to recover by stammering "or anyone else's for that matter." The rest of us broke up laughing. Fortunately, so did Auerbach.

Dr. Baker turned out to be a perfect professor. He was helpful, friendly and always there with advice. But he trusted us to show our own initiative and to set our own agenda. The first month I was in Chicago, I saw Baker come rushing up from the departmental office with his mail. "Oh boy," he chortled, "the new issue of *Genetics* has come in." And with that, he tore open the envelope, plopped down at his desk and started to read with relish.

I was astounded because reading a scientific paper was agony for me. I painstakingly read lines over and over again in order to understand. I couldn't, for the life of me, see how anyone could actually *enjoy* reading a paper. What I didn't realize then as a green graduate student was that experience with the terminology and the techniques would eventually make that dense jargon perfectly understandable. Moreover, one learns to skim. It's possible to check out the Summary, for example, and see what the main conclusions are. If they look interesting, then one might read the Introduction, the Results and the Discussion. The details of the experimental procedure are not that important unless one wants to try some of the experiments or check on some specific point. Years later, I was every bit as excited and pleased as Baker when one of my favourite journals arrived in the mail.

It's hard to believe that I have now reached the age Baker was when he was my professor. To me, he seemed so old and wise, while I feel I'm still young and know I'm ignorant. He was a good sport and I talked him into going to the gym regularly to play basketball. I had always been taught to treat older people with respect, so I found it hard to address him with familiarity. It took me a long time before I could call him "Bill." When we played in pick-up games I felt silly yelling, "Pass the ball, Dr. Baker." It's a rare quality when a professor can be a good friend, yet still maintain sufficient distance to remain a mentor. Baker did this very well.

While I was in graduate school, the one-hundredth anniversary of Darwin's famous publication, *The Origin of Species,* was celebrated in 1959 with a special symposium at the University of Chicago. It was an impressive event with great scientists coming from around the world. One of them was H.J. Muller. During one session, someone asked whether life would ever be created in the lab. Different panel members talked about accomplishing this in a few years, decades or centuries. Then Muller got up and electrified us by saying it had *already been done* through the test-tube replication of DNA. Few people would have agreed with his definition of "life," but he gave us a perspective on how close we really were with our manipulation of the foundations of life itself.

After one of the sessions, I got on a bus to go home and found myself sitting behind Muller and his wife. I leaned forward eagerly to eavesdrop on their conversation, expecting to hear some profound observations on one of the lectures. Instead, I heard his wife admonishing him for having his tie on crooked! It was an important insight for me — to realize that even gods have to deal with life's more simple things.

My interests in civil rights continued to grow while I was in graduate school. In Chicago, I joined the Japanese-American Citizens' League (JACL), the U.S. counterpart to the JCCA. I played in a basketball league of Japanese-Americans. My other motivation for joining the JACL was an extension of what I had felt in London, a feeling of social acceptance among people with a shared set of experiences. I even got a small grant from the JACL to conduct a poll of several hundred Japanese-American teenagers. It was an attempt to document the degree of their assimilation and the persistence of Japanese customs, such as language, food and culture. My survey was crude, but revealing. Japanese-American teenagers in the sixties were academic achievers and considered themselves full Americans. Yet in eating habits, cultural interests and friends, they were clearly not in the mainstream of white society. That has all changed enormously in recent decades.

In Chicago, racial issues revolved around the black community. To me blacks were always a symbol of all oppressed groups because they were such palpable victims. My father had always spoken of black people as a kindred group — their fight was ours. I saw my hang-ups about being Japanese in a white society as a mini-reflection of black problems.

The University of Chicago lies right along the edge of the "Black Belt," on Chicago's south side. The neighbourhood where most students lived was rundown and crime-ridden, and we walked a constant gauntlet of potential muggers. Few of us in the department escaped from Chicago without being assaulted or robbed. I was appalled by the violence and *de facto* segregation. In Chicago, there were restaurants that didn't serve blacks in the fifties.

In 1960, four black students created a new way of protest in Greensboro, North Carolina. They sat down at a segregated lunch counter in Woolworth's, thereby creating a "sit-in." This tactic was copied across the country. In that year, SNCC (Student Non-violent Co-ordinating Committee) was founded and eventually charismatic people like Rap Brown and Stokely Carmichael emerged to lead the organization.

On May 4, 1961, James Farmer, head of CORE (Congress of Racial Equality), left Washington, D.C. on a bus. It was the start of the famous Freedom Ride which met violence when the integrated bus hit the Deep South. Vicariously, I rooted for the riders, seeing in their victories the defeat of bigotry — our common enemy.

My research was advancing well. Because I had decided on a thesis project soon after I arrived at Chicago, I was able to start accumulating data right away. I graduated in June 1961, less than three years after I had arrived with my B.A. When I received my Ph.D., it was handed to me by the brand-new president of the University of Chicago, George Beadle. Beadle was a geneticist who had earned the Nobel Prize for his work showing that genes control the production of proteins. When I accepted my degree from him, I said, "Thank you, fellow geneticist." In retrospect, it seems like a pretty cocky thing to say, but he rewarded me with a big grin and a handshake.

After I finished my thesis, I applied to present a paper on my thesis work at the annual meeting of the Genetics Society of America at Corvallis, Oregon. I had come up with a hypothesis to explain my results and I wanted to try it out. I was accepted. The talks were short fifteen-minute presentations with questions from the audience. This was my professional debut as a fully accredited geneticist. I gave my presentation shaking in my boots because Jack Schultz, one of the giants in my area, was in the audience.

As I feared, when I finished, Schultz got up immediately. He looked at me and shook his head. "Dave, Dave, Dave," he started, "you've gone wild. You've gone completely overboard." He then launched into a point-by-point critique of my work, referring generously to his own, of course. I was demolished. One of the gods had pronounced judgement on my work. But on reflection, I realized that he was just defending his turf. He had been in the field for decades and, by god, he would slap down any newcomer. It was all part of the game. Because of some of Schultz's criticism, I delayed publishing my thesis results and repeated all my work, as he had suggested. It paid off because I did get some different results. However, I never changed my hypothesis, and Jack and I became scientific friends.

TO OAK RIDGE, HOME OF THE BOMB

During one of the "Midwest *Drosophila*" meetings at Madison, Wisconsin, I met Dan Lindsley, another scientific titan whose enormous reputation as a chromosome mechanic was richly deserved. He was visiting from his lab at Oak Ridge, Tennessee. You may recognize the name Oak Ridge in association with the Manhattan Project, as the site to purify uranium for the atomic bomb. It had been chosen because it was away from urban centres and sufficiently isolated to allow secrecy. It has lots of rolling hills to absorb and disperse any large explosion. And there was plenty of cheap energy from the system of dams and waterways of the Tennessee Valley Authority. After the war, the Biology Division had been set up, ostensibly to study the biological effects of radiation. But under the leadership of its director, Alexander Hollaender, it became one of the leading centres in basic genetics. Dan Lindsley headed up an outstanding group of fly geneticists who had a common skill in chromosome mechanics.

Professor Baker had worked at Oak Ridge himself and urged me to apply for the position of Research Associate once I completed my Ph.D. It was like a paid job. I would be an employee of Union Carbide — now infamous for the Bhopal tragedy in which a deadly gas escaped in India, killing and blinding thousands of people — but at Oak Ridge, the company simply administered government money. Hollaender made all of the decisions. People like Lindsley were full-time members of the Biology Division, while Research Associates came for varying periods as postdoctoral fellows to hone research skills.

It did not take much convincing for me to try for a position at Oak Ridge. I felt excited and privileged to be considered for a post with Lindsley. He was extremely low key and approachable. He didn't make me feel inadequate at all, even though I knew he was in a completely different league.

I was extremely nervous about living in the South. My identification with blacks by this time was almost total and I didn't know how I would react to living in an area of *de facto* segregation. The labs were all completely integrated and on my visit, all appeared harmonious. In fact, as I later found, the bigotry was not far below the surface. But my enthusiasm for the scientists overcame any other concerns. And after flying there for a personal interview, I was offered a post.

Looking back on the year I spent at Oak Ridge, I see it as a scientifically idyllic time. I had received my Ph.D., certifying that I was a fully licensed scientist. I had arrived in Dan's lab free to do whatever I wanted. I didn't have to sit on committees or do administrative work. There has never been a time since when I've been so free.

When I got to Oak Ridge, one of the first formalities was to talk to the fly group about my thesis. When I finished, Lindsley outlined a particular experiment he thought I should do. I couldn't, for the life of me, see what his suggestions had to do with my ideas. After almost a year, repeating, cleaning up and extending my thesis work for publication, I realized that the final critical test had been the experiment he had suggested when I first arrived!

There was a wonderful atmosphere in the Biology Division at Oak Ridge that I've never experienced anywhere else. You could go back at night or on weekends and always find people there hard at work. The building itself was terrible! The fly lab was buried in the bowels of a huge building. It had no windows anywhere and there was always the racket of the air conditioning which kept the temperature constant. But people were excited about their work, and would often visit each other at any time.

They would drop whatever they were doing and talk for hours. People were always collaborating on ideas and projects. There were seminars and "journal clubs" to report on current papers going on every day to allow us to keep abreast of research around the world. Those of us who had attended meetings would give reports on what we had learned. Looking back at the roster of people there, I'm astounded at the collection of talent.

I took advantage of the freedom and cross-pollination of areas. Just for the hell of it, I thought there might be ways to test out some of my ideas on chromosome behaviour in *Drosophila* in other organisms. So I hung around the lab of Fred deSerres who worked with the bread mould, *Neurospora.* I thought it might be interesting to try some chromosome mechanics on an organism that could be handled like bacteria or viruses. I also thought it might be fun to *look* at some chromosomes, so I got Drew Schwartz to set up crosses for me with *corn,* which have large chromosomes. The division thrived on such informal collaboration.

It was during my time at Oak Ridge that Salvador Luria, the MIT virus geneticist who would later win a Nobel Prize for his work, came to the Biology Division to give a lecture. After being introduced, he rose and committed a terrible *faux pas* from the standpoint of his own work. He started off by saying, "I don't know why you're all interested in hearing about *my* work, when Francis Crick has solved the whole question of the genetic code." It was electrifying for the audience! He was referring to a paper that hadn't yet been published and so was completely new to us. He then proceeded to talk for an hour about his own work but all of us were wondering: "What has Crick done?" As soon as Luria finished his talk, that was the first question asked. There followed one of those rare moments in science when we became privy to a truly profound insight.

Now remember, we knew genes are DNA and that the sequence of the four different bases along its length is like a message. We could think of a gene as a sentence containing information about how to build certain important molecules. But how is the gene read? Luria proceeded to describe how Crick and his associates had chemically scrambled a specific gene. In a series of brilliant manoeuvres, they were able to show that a gene is "read" from one end at a starting point that functions like the capital letter in a sentence. A cell then proceeds to read the gene's information in a fixed "reading frame" of three letters at a time. The genetic code is written in triplets.

Luria carried a "preprint" of this paper with him and he gave it out to be photocopied. The audience went into a frenzy of excitement and we pored over the manuscript, aware that we were reading a "classic" piece of work. And it was. I still have a scruffy copy of that preprint.

The article actually came out in December 1961 in the eminent journal, *Nature.* The paper marked the beginning of a period when most of the details of what is now molecular dogma were uncovered. In later years, biochemists would carry out work entirely confirming Crick's conclusions. What made us proud was that Crick had done it all by using *genetic* analysis.

Oak Ridge was a government town, owing its existence to the facilities to extract radioactive uranium. There were a lot of Northerners living there and that gave the impression the town was an oasis of liberality in a sea of rednecks. But it was still the South. No blacks lived within the city limits. Blacks weren't allowed to use the coin laundromats, swim in the public pools or go to the drive-in theaters. I joined the NAACP and was the only non-black member.

Ruby Wilkerson was a black who was the stock-keeper in our lab. She was a wonderful woman, deeply religious, always cheerful and blessed with a marvellous sense of humour. She and her husband, Floyd, who worked in the facilities for mouse genetics, lived in Philadelphia, a small rural town about thirty miles from Oak Ridge. Ruby and I became so close, we were like siblings. My wife and I would go to her place for wonderful family dinners. When we arrived, it was an event and the entire Wilkerson clan would show up to share an immense banquet.

During the dinners, the Wilkersons would leave the television set on for the kids to watch. I remember once I was talking and suddenly realized that no one was paying attention to me anymore. Everyone was glued to the TV. I turned around to look and there was a black person on the screen. The family was starved for black faces to give them a connection with what they were watching. It was a striking example of the importance of role models with whom people can identify. Sometimes I wonder: What do Native people in the high Arctic think as they watch "Dynasty," "The A-Team" or "Seeing Things"?

Blacks I met in Oak Ridge were like time bombs waiting to go off. Outwardly, they expressed friendliness and a kind of simpleness that was at once self-demeaning and yet mocking. But in the security of their homes, smouldering resentment and frustration would flame up.

I found myself identifying as a surrogate black. Their resentments became mine. Once one of the white handymen in the lab, a man from Georgia, offered to fix my car. He came over on a Saturday and proceeded to yank my engine apart and grind my valves for me. During the day, we did a lot of chit-chatting, and at one point he informed me that the notion of racial intermarriage made him physically ill. Gradually he revealed an attitude of bigotry towards blacks that took my breath away. He had a contempt for them as human beings, a belief that they were, in fact, inferior and stupid. And I, with the selfish reaction of concern for my car, let him go on without objecting. How I have regretted that I didn't have the courage to tell him to go to hell on the spot. When he left, I was limp with rage.

Dad and mom came down to Oak Ridge to visit. The many lakes and rivers of the Tennessee Valley Authority power grid were full of trout, striped bass and shad. Dad bought a fishing license and was surprised to see a space for "colour." He puzzled over it and finally wrote down "brown."

My wife and I bought a large station wagon and we took trips into the Blue Ridge Mountains to camp and fish. On long weekends, we went on forays — first to Knoxville, Chattanooga and Memphis. Then we made two longer trips into the Deep South. I wanted to see what I had read so much about. We went to Little Rock, Arkansas, where Governor Orval Faubus had fought integration of the schools, and I was thrilled to drive by a high school and see black and white students mingling nonchalantly.

We drove through Montgomery, Alabama, the largest industrial city in the South and then still rigidly segregated. In 1963, Eugene "Bull" Connor would make international headlines with his brutal assault on demonstrators. We were very nervous when we pulled into a motel for whites. But there seemed to be no discomfort with our foreign appearance and accent. It was on one of these trips that I began to feel a rage towards white people. A "Whites Only" sign drove me into a frenzy of anger.

Racism grinds the victim down. It's the accumulation of little daily insults to one's dignity — having to step aside for a white person on a sidewalk, being looked right through by clerks as if one is invisible, recognizing the nuance of distaste or superiority on a face. It can break the soul or ignite a fire in it. Gradually, I found I couldn't stand the drawl of a southern accent. I assumed that anyone with white skin was probably a bigot.

Meanwhile, I had completed most of the experiments to round out my thesis work and had published two papers. I decided to apply for a number of positions. Frank Ratty was on sabbatical at Oak Ridge. He was a fly geneticist and the Head of the Biology Department at San Diego State University. Frank urged me to apply for a job in his department. I did and was offered a job. It's a wonderful part of the country and I was very tempted. I also applied to Stanford to work with Dave Perkins, who was a *Neurospora* geneticist. He offered me a fellowship. My first choice was to spend a year with the eminent fly geneticist, Mel Green, at the University of California at Davis, and he quickly offered me a position. Scientifically, this would have been a great move. But by then I was consumed with bitterness and anger at the racism apparent all around me. I finally decided that I had to leave the United States altogether and return to Canada.

I knew that I was returning to a country which had incarcerated me as a Japanese-Canadian and which treated Natives as badly as Americans treated blacks in the South. But Canada was smaller, and I felt there would be more opportunity for an individual to register an impact there than in a huge country like the United States. To me, Canada still meant a civility that was missing in the States. Canada was Tommy Douglas, medicare, Quebec, the NFB and CBC, differences that mattered, not because Canadians were better, but because they were different and, for me, preferable. From the standpoint of my career in science, the choice was stupid. But I have never regretted my decision to come home.

I applied to the National Research Council of Canada for a postdoctoral fellowship, but was turned down. At that time, the awards were given mainly to graduates of Canadian universities. I went to the Genetics Society of Canada meetings in Winnipeg so I could look for a job. There I met Clayton Person who had recently been appointed head of a new Genetics Department at the University of Alberta in Edmonton. On the basis of my scientific presentation and our conversation, he offered me a job. He told me years later that when he discussed me at a faculty meeting, everyone had been very positive and one of the senior people had said, "Yeah, he seems good. And he doesn't even *look* Japanese." If his genetics was as good as his eyesight, he must have been a pretty poor scientist.

Once again, I had reached a threshold. My scientific training complete, my field of interest now set, I was going home to Canada after spending eight years in the United States to start a lab and to see how well I could do on my own. Gone was my original limited goal of being a teacher, and enjoying genetics vicariously. I had worked with the best and believed I could compete with them. I was a full-fledged scientist and wasn't afraid to work hard and long.

In 1958 Marcia marries Dick Aoki. Seated (l to r): Dawn, grandmother Suzuki, mom, Dick's mother. Standing: Aiko (2nd left), grandfather Suzuki (4th left), me (back), Marcia, Dick, Dad (2nd right), Dick's dad (right).

3

Mom and Dad on their 25th wedding anniversary.

4

(above) August 1958, Newlyweds. Joane and me.

(above right) Joane and Tamiko on the dunes of Lake Michigan.

(right) Dad visited us in Oak Ridge. This is Tami at two.

My lab bench area in Oak Ridge in 1962. On the left is Dan Lindsley, the head of *Drosophila* group. On the right is Yataro Tazima, one of the world's leading experts in the genetics of silkworm moths in Japan.

Scientific meeting in Gatlinberg, Tennessee in 1962. These are some of the Oak Ridge scientists. (l to r): Charlie Mead, David Krieg, "Tex" Barnett, me, G.D. Novelli, F. Sawada

CHAPTER SIX

OBSESSION — SCIENCE AND RESEARCH

GENETICS BECAME an all-consuming passion for me in my last year at Amherst and remained so for almost two decades. It overwhelmed the rest of my life. Marriage, homelife, friends — all had to accommodate to my need to "go back to the lab." I threw myself into research seven days a week, day and night. For me it wasn't work; it was total involvement. There was nothing else I preferred to do with my time, and I used to marvel that I was being paid a good salary for a pursuit which I'd have done gladly for nothing.

After my children were born, I devoted as much time as I could to them; even after my marriage broke up, I saw them every day. But all else was overridden by the lab. The lab was my work, my social life, my extracurricular activity. The students and technicians who worked with me became my extended family. We drank and camped together, played football and spent long nights "pushing flies."

Joane and I were twenty-two years old when we married, right after I graduated from Amherst. Our eldest daughter, Tamiko, was married in 1984, two years after graduating from McGill University. She was twenty-four, and at the wedding I leaned over and whispered to Joane, "They're so young!" Joane reminded me that they were not only two years older than we had been, but also more sophisticated. I, after all, had married my high school sweetheart, the first woman in my life.

I don't remember any strong pressure to get married; it was just something we expected to do. From the time we had graduated from high school, the people we knew started to get married. For those of us who went on to continue our studies, college graduation was the point at which lots of us got married; it was expected. For me, it made economic sense because while my stipend at Chicago was sufficient to live on, our combined incomes would enable us to live more comfortably.

During our high school years, Joane was a wonderful companion and a great sport. When we started going steady in grade twelve, we did all of the things that teenagers do. In hindsight, I can see that we didn't have the base of shared interests and dreams that sustains a long relationship. But back then, it didn't seem to matter. Like so many teenaged girls, Joane was willing to let my interests and ambitions dominate her own.

Joane and I had our lives planned out to the end of graduate school. She would work as a technician in electron microscopy while I worked on my Ph.D. We could live on her salary and stash away whatever I made in scholarships and teaching assistantships. Children weren't in our plans until long after I got a permanent job. Within five months of marriage, Joane was pregnant.

I had never grown up around young children and had never even given a thought to having a family. So my impending fatherhood was a shock that filled me with fear. I had no idea what to expect or do. We dutifully took a course on how to care for the baby once it was born, but in those days fathers were not expected to participate in the delivery process. Birth in the fifties was a "managed" affair, with doctors in charge and the mother treated as if she were ill. I waited outside the delivery room with other expectant fathers.

On January 21, 1960, Tamiko Lynda was born. Tami's arrival, while so disruptive and unexpected, opened a wonderful dimension of parenthood in me. Joane went back to work soon after the baby was born, at first half-time and then later full-time. Since

I was well into my thesis project at the University of Chicago and had completed most of my course work, I was able to adjust my time to take care of Tami while Joane was at work. I loved to study late at night with Tami lying asleep on her tummy on my chest. I had long developed the habit of working late into the night anyway, so when Joane came home from work, I would often be heading out the door to put in a night at the lab. On occasion, I would also take Tami to the lab and park her buggy next to my microscope while I counted flies.

I had never experienced the "giving" part of love until Tami came. I became a typical crazy parent and whipped out photographs of Tami at every opportunity. Today I love to watch young couples who have just had their first baby. They act just as I felt, as if they had made some Nobel Prize-winning discovery.

I helped take care of Tami, bathing and feeding her and changing diapers. I've never understood why changing diapers is considered such an onerous task. Japanese people aren't hung up about body functions. To me, changing diapers is one of those intimate experiences that bonds us to babies. Whenever I changed my children, then picked up the soiled diaper, I always felt very moved by the *warmth* radiating out of the soggy cloth. That heat had come from my child's body, and as the diaper cooled, it was to me somehow symbolic of how transitory our lives are here on earth.

Tami became the focus of our lives and made us want more children, even though the baby had taken us from living very comfortably on Joane's income to living frugally on our combined salaries. This gave me a lot of incentive to finish my studies quickly. And I did — in less than three years. When we arrived at Oak Ridge, Tennessee, for my postdoctoral year, Joane was already pregnant again, this time deliberately. And I was very enthusiastic about doing research.

Troy was born almost exactly two years after Tamiko. By the time he arrived, I was well into my research and thinking about getting a permanent position. Tami was continually surprising us with each new step in her development, and as he grew Troy was charting now-familiar territory. I think it must be the bane of all second born. But he was my father's first male grandchild, and therefore, in the continuing pattern of Japanese thinking, very special. Troy was a wonderful child, good-natured and undemanding; it's ironic that had he been a problem child, I might have given him more time.

Joane was a marvellous mother and a good wife. During those times, she bore my long nights and weekends at the lab and never complained. I called the shots — deciding when and where we'd take a holiday, when I'd come home for dinner, whom we'd have over. And all the while I couldn't share with my wife the excitement of what was going on in the lab. I'm sure that when I said "I'm going to the lab" to Tami and Troy, they imagined me disappearing into some dark and mysterious hole.

THE PRODIGAL RETURNS TO CANADA

We left Oak Ridge for Edmonton, where I took up my first faculty position. I had to set up a new lab, start my experiments, teach a new course, and apply for a research grant. If Oak Ridge had taken a lot of my time, it was nothing compared to the University of Alberta. I hurled myself into my teaching and found my social life revolving around my students. This drew me further away from home life. I don't even know now what Joane did with the children, especially during that long and cold winter in Edmonton. When I came home for dinner, the children were there for me to play with, bathe, read to and tuck in bed. The family seemed to be there waiting for my whim and inclination.

I had accepted the job at Edmonton in order to return to Canada, away from the United States where the magnitude of racial inequity had overwhelmed me. But I had not realized how far north Edmonton was. It was a raw city, vibrating with energy, but saddled with what I thought were enormous handicaps — a flat landscape relieved only by the river, mosquitoes as big as sparrows in the summer, and winters that were inhumanly cold.

We reached Edmonton just prior to the fall semester of 1962, having driven and camped across the southern United States to

California, up the coast and then inland to Alberta. We lived on the University of Alberta campus in faculty quarters, only three blocks from the Agriculture Building where the Genetics Department was located. In the dead of winter, I couldn't make it all the way to the lab without ducking into another building to warm up.

While Joane was kept busy with two young children, I plunged into my new job. My major concern was to establish facilities to grow flies on a sufficiently large scale to enable me to start my experiments. I had to apply for a research grant which would let me buy equipment, hire a technician and pay students for summer work. I was soon to discover the differences between Canada and the United States. My American peers, starting out as assistant professors like me, could expect their first grants in the $30,000 to $40,000 range. I was told that National Research Council of Canada grants began at about $2,500!

The Department of Genetics at Alberta had just been established; it was the second in all of Canada (the first being at McGill). As part of the Agriculture Faculty, it was heavily laden with animal and plant breeders. These were people who either worked right on the farm breeding animals or selecting stocks of wheat, or they were theoreticians who studied the mathematics of genes in populations. Either way, I wasn't interested — meaning I didn't know anything about these areas. Molecular genetics was where the action was for me. There was a predictable tension between the more "classical" geneticists and those, like me, who were excited by the developments of Watson and Crick.

The department had some good faculty, but none with that burning drive to produce. At night, my lab was the only one lit up and active. That constant exchange of ideas and excitement over new papers that I had previously experienced at the University of Chicago and at Oak Ridge just wasn't there. I was considered aggressive and nakedly ambitious — "brash" was the word I frequently heard.

Clayton Person had been brought in as the man to get the department started, and he was perfect for the job. He was himself classically trained and had made his mark by proposing an idea of great importance to breeders. Clayton was also keenly interested in molecular genetics. He had instantly recognized the significance of the Watson-Crick DNA model when it came out in 1953 and had published a paper showing how the molecule could replicate. So he straddled both the classical and modern phases of genetics. But Person's most important trait was his personality. He was rock solid in integrity and there wasn't a devious bone in him; he was respected and loved by all. I admired him immensely. He realized that if the Genetics Department remained in Agriculture, genetics would be held back in the fundamental areas. So he fought to move the department to the Faculty of Science and succeeded.

During my first semester at the University of Alberta, I had no formal teaching assignments. Instead, I was able to devote all of my time to setting up a lab and preparing lecture notes for my course in the second semester. The department supported a staff of people to wash equipment and paid for the material I needed to get started. My initial needs were simple: glassware, fly food and temperature-controlled incubators. Even back then, Alberta was, by Canadian standards, enlightened and generous towards its scientists. We received much more support from the provincial government than I have ever seen in British Columbia or most of the other provinces.

The downside was the Edmonton winter where the temperature plummeted to minus forty degrees Fahrenheit. In the middle of that winter, a job was advertised at the University of British Columbia. I felt guilty about considering another job so soon after I had arrived in Edmonton, but the cold and the fact that Vancouver was a big part of my childhood folklore made the offer attractive. I applied, and as a prelude to being considered was invited to give a seminar at UBC. When I left Edmonton, the temperature was minus thirty degrees. When I arrived in Vancouver, it was plus thirty degrees and everyone was complaining because it was so cold! Vancouver was warm and breathtakingly beautiful, but what sold me were names of places that I'd heard my parents use so often — Granville Street, Grouse Mountain, English Bay, Kitsilano, Stanley Park. It was thrilling for me to have a physical reality with which to match those names.

I gave my talk and it was well received. I wanted the job and said so to the people who were my hosts, but this was not the way to bargain for a job. Perhaps that's why I was given an offer at a salary several hundred dollars below my University of Alberta wage. I took it anyway. (I had been so anxious to return to Canada that I took a pay cut from my Oak Ridge salary to go to Edmonton. At this rate, I was heading back to my graduate student stipend.) I felt terrible about leaving Person after all he had done for me, and I told him that if he asked me to stay in Edmonton as a special favour, I would. But I knew he wouldn't ask; he had too much integrity. I was delighted when, a few years later, he also left Alberta to join the Botany Department at the University of British Columbia.

When I called my dad to tell him I was accepting the job at UBC, the first thing he blurted out was, "They kicked us out of there twenty years ago!" After we had moved to Ontario, dad never talked about the war years, yet that was his immediate reaction two decades later.

I had spent only a year in Alberta, but it was invaluable. During my time there, I had adjusted to the realities of the Canadian scientific establishment and Canadian funding, met a few people, got my own lab started and taught my first course. In the spring of 1963, I received a grant from the National Research Council for $3,600, larger than usual because of my experience at Oak Ridge.

UBC AND TEMPERATURE-SENSITIVE MUTATIONS

When I left for Vancouver in August of 1963, I had completed all of the experiments I needed to publish the work I had started at Oak Ridge. Once I got to Vancouver, I was ready to begin fresh work that was completely my own.

Science is a game for young people. They are up-to-date on what's going on; they have the enthusiasm and the energy to pour themselves into it. At the University of Alberta, I only had to teach one semester course for the entire year, so I had plenty of time for research. In contrast, at UBC the younger members of the faculty were loaded with teaching, while older, established people taught the smaller, more specialized courses in the upper years. This is not the right way to get the most out of young scientists.

Despite the teaching load, I managed to cloister myself in my lab and my research. I was especially interested in chromosome behaviour during cell division, particularly the way chromosomes paired and exchanged parts. My work at UBC led inexorably to the conclusion that the complex process of cell division, with its elegant steps of chromosome duplication, pairing, crossing over and separation, must be under rigid genetic control. In theory, it should be possible to detect mutations in such genes that would disrupt some of these events. One should be able to dissect the steps in cell division through the detection of a whole series of such defects.

A series of fruitfly experiments led me to suggest that there were cell division genes which were so important, they were present in multiple copies or duplicates. This meant that a recessive mutation couldn't be observed because it would be masked by the normal dominant duplicates. A dominant mutation, on the other hand, would kill or sterilize the fly if it disrupted as important a process as division. So I was stuck. How could I induce and recover dominant mutations if the flies that carried them were dead or sterile?

In the summer of 1965, I attended a meeting of the Genetics Society of America in Fort Collins, Colorado. There I met a bacterial geneticist from Berkeley, named Joe Clark. We hit it off right away. One night we were at a bar together and I told him about my scientific dilemma. As soon as I defined it — he said "temperature-sensitives." The instant he said it, I knew he was right. I returned to my office excited at the idea of looking for dominant lethal mutations that were temperature-sensitive in *Drosophila*.

Microbial geneticists had long ago discovered that it was possible to pick up mutations that were lethal under certain conditions, but perfectly healthy under a different set of conditions. They're called "conditional" mutants because their survival or death depends on the environmental conditions. Temperature was one of the most commonly used conditions.

Temperature-sensitive (ts) mutants grow well at one temperature but fail to survive at another. It had been shown in bacteria and viruses that temperature sensitivity was caused by a change of an amino acid in a protein. At one temperature, that mutant protein could attain its proper shape to function, whereas at another temperature it would not be stable enough to maintain the normal configuration, so the protein function would be lost.

By this time, I had a top-notch scientist to manage the day-to-day operation of my lab. She was Leonie Piternick, who had earned her Ph.D. at Berkeley under another great German geneticist, Richard Goldschmidt. Leonie had come to Vancouver with her husband. One day, she appeared in my lab and volunteered to push flies for me. She loved *Drosophila* genetics and worked long hours for next to no money. It was the worst kind of exploitation, but I had no grant money. I was very grateful for her help.

Temperature-sensitive mutations had proved extremely useful in microbial genetics, but the reaction of *Drosophila* people (and my own initial inclination) to my proposal to screen for them was, "It's one thing to get them in micro-organisms, but it would be too difficult to find them in a complex creature like *Drosophila.*" We were helped by the fact that powerful chemicals existed that induced a high frequency of mutations. The chemical we used, for example, attacked one of the four bases of DNA to cause an alteration from a G-C base pair to A-T. Just by feeding the compound to males, we could induce mutations on just about every chromosome carried by each sperm. Thus, using chemicals, it was feasible to look for rare kinds of events.

We decided on a very simple protocol. We would expose male

sperm to a chemical mutagen, then "clone" each treated X chromosome so it could be tested at two different temperatures. We looked for mutations that caused death at 29°C but survived at 22°C. Before we set off to find dominant lethals which had never been studied before, we decided we'd better first determine whether recessive ts lethals could be recovered. If we couldn't find them, there would be no point in searching for dominants. The procedure we adopted was pretty laborious, so Leonie and I set an upper limit of five thousand cloned chromosomes. If among that many we didn't find any ts lethals, we would give up. That, we figured, might take a year.

Within weeks, Leonie had found several ts mutations in the first hundred. We had struck a motherlode! So long as we grew flies carrying the mutations at 22°C, they appeared to be perfectly healthy and fertile, but when they were raised in an environment only seven degrees warmer, they died. One of the first things Leonie did was to put 22°C-reared adults carrying a ts lethal mutation into the 29°C incubator to see how long it took before they died. To her surprise, they survived, laid eggs and often the eggs hatched. But how could this be when they were carrying a mutation that killed them at that temperature? We immediately realized that the mutation must no longer be active in the adult, that it functioned at some earlier stage in development. Thus, ts mutations held possibilities for probing stages in *Drosophila* development.

What geneticists do is comparable to a surgeon removing an organ or tissue, or a mechanic taking out a component of an engine, and then studying the consequences. Our scalpel or wrench is the mutation. A mutation is a very subtle way of knocking out a cellular component. Often it has a cascade effect; its absence knocks out other systems until death or a defect results. With ts mutations, we could refine this probe even further by pinpointing the action of the gene to a specific time in development and even to a particular part in the body.

Leonie's great fascination was with mechanisms controlling development and, thanks to Baker in graduate school and Schotté at Amherst, I had always been interested in similar questions. Now ts mutations gave us a wonderful tool to pursue our interests. We found ways to synchronize the development of flies so in any culture they were all at the same stage. Then we would give different vials a "heat shock" at different stages. If the flies died when exposed to the killing temperature, we would delineate the actual ts period (tsp), which we assumed was the developmental interval when the mutant gene was active or its product needed.

Soon we had dozens of different ts lethals, and we put our search for dominant ts lethals on hold so we could exploit the material at hand. Each mutation gave new and interesting results. One strain had a series of successive tsps; that is, it was sensitive to high temperature, then insensitive, then sensitive again, and so on.

While giving a lecture at Berkeley, I mentioned these results and suggested that an explanation might be that a gene is turned on and off repeatedly during development. In the audience was my great hero and scientific grandfather, Curt Stern. He rose at the end of my talk and proposed that another interpretation might be given. "Perhaps," he said, "the gene is turned on and off in one tissue, then on and off in another tissue and so on. So in the life of any cell, the gene would only be activated and shut off once." Years later, when we found a mutation that could be tested in several tissues, we found that Stern had been right. That kind of immediate insight comes from long experience.

Once we had announced our findings in a prestigious journal, there was a great deal of interest from microbial and molecular geneticists who immediately recognized the value of ts mutations for probing development in a more precise way. They, after all, had been the ones to exploit ts mutations in the first place. But my fellow fly geneticists were another matter. They were much more skeptical of the value of our mutations. To me, the utility of ts lethals to study *Drosophila* development was beyond doubt. But many fly people wanted to know the molecular basis of temperature sensitivity, because that affected the way we interpreted the results. It was well known that in micro-organisms a change in a single base pair of DNA was enough to alter the properties of the gene product and render it thermally unstable. We assumed a similar mechanism for our mutations, but we had no direct evidence.

I was invited by a number of universities to lecture on our work. While at Rockefeller University, I met Ed Reich, a Canadian on faculty. Ed was famous for having determined the molecular effects of a potent antibiotic called actinomycin D, and since I had worked with the drug and knew of his work, I was thrilled to meet him. He was fascinated with our experiments, and when I told him about the need to determine the molecular basis of temperature sensitivity, he thought out loud: "You need a mutant that affects a common protein in the fly. Then you can get enough material to work with. The best candidate should be muscle. If you could get a ts muscle mutant, that would be fascinating. Look, I'll do the biochemical analysis for you."

It was an exciting suggestion and an offer I couldn't refuse. I returned to UBC all fired up to look for a muscle mutant. But how would we recognize it? I spent long hours talking the question over with people in the lab. Two students, Rodney Williamson and Tom Grigliatti, were especially keen. Rod had been born in England, grew up in Pasadena and somehow ended up at UBC. I think he drifted into our lab as much out of interest in one of my female students as in genetics. But he had a keen mind and a dogged determination once he got going. Tom was a Californian who had come to work on his Ph.D. with me after getting a Master's degree from San Francisco State. He loved the challenge of new ideas and was forever doing experiments unrelated to his thesis "just for the hell of it." The two of them were dynamos, matching me for time spent in the lab seven days a week.

We soon developed a plan. We would hunt for ts muscle mutants by screening for adult flies that were paralyzed at high temperatures and which recovered mobility when returned to lower temperatures. Looking back, a search for a reversibly paralyzed adult fly seems like a crazy idea, but we were undaunted.

At a meeting of *Drosophila* geneticists, when I mentioned what I was going to do, everyone thought I was crazy. And here is one of the few advantages of working in the boondocks like Canada. We aren't under the same kind of pressure to produce as the big-name American labs, so we can afford to take a chance on a nutty idea. As well, Canadian funding agencies have a good policy. Once grant money is committed, scientists are relatively free to use that money for research as they see fit. That means if something unexpected suddenly comes up, we can divert resources into it without first having to go through a lot of paperwork to justify the decision to some grant committee.

We knew we were looking for something that would be extremely rare, if it occurred at all. So we had to be able to collect large numbers of flies to treat with chemical mutagens, and then to rear them in huge cages. We invented a way to use ts lethals to give us *unisexual* progeny, either males or females, thereby saving the time spent in separating the sexes. Rod built a contraption that let him isolate a few paralyzed flies from thousands of active ones, without releasing a cloud of escapees into the lab. And we scaled up our rearing cultures so we could collect tens of thousands of flies quickly.

All of this took a lot of time to work out, but *that* is the fun part of science. I'm disappointed at how science is often presented in schools. Youngsters may come into a science class to be handed a sheet of instructions and materials and are then told to "do the experiment." Frequently, they don't understand what is going on but they dutifully follow the recipe. And too often the most important exercise is to write up the results *properly* — Purpose, Methods and Materials, Results, etc. This concept of science turns kids off. I'm disturbed because that is *not* what science is. Science as I know it is getting excited over that crazy idea of looking for paralyzed flies, trying to figure out how to do it, playing around with equipment and stocks, and then spending all that time doing the experiments. It's only *after* this involved procedure that we examine the data and finally write up a paper. The research report is a way of "making sense" of what has been done, putting the work into context, communicating results and drawing conclusions. But that's done long after the exciting research has been carried out.

After months of grinding through lots of flies, Rod and Tom called me over to their lab bench. "Have a look at this," Rod told me, and he held up a vial of flies. Running and flying around, the flies inside looked fine to me. Rod then wrapped his palm around the vial and held it for a few minutes to warm it up. When he opened his hand, all of the males were lying motionless on the bottom while the females were still running about. When Rod shook the flies into a cooler vial, the males were already flying before they hit the bottom. These males were carrying a kind of mutation never before known. After looking through over a quarter of a million flies, Rod and Tom had isolated the world's first temperature-sensitive paralytic mutation.

Words can't convey the thrill of that moment. I screamed, danced up and down, and then shook like a leaf for hours. We knew that it was something completely novel and momentous. The fact that people had been so negative when we started only added to the intensity of our pleasure.

What was the mutation doing? The speed with which it responded to temperature change suggested that the mutation could just as readily be a defect in nerves as it might be in muscle. And it was. By screening for paralysis, we had unexpectedly found a way to recover nerve defects. We named the mutation *paralytic temperature-sensitive* ($para^{ts}$), and it led us into a completely new area of study—the genetics of behaviour and neurophysiology.

Eventually, after screening over one and a quarter million flies, we recovered several different ts paralytic flies and a veritable zoo of other strange and wonderful nerve defects. In the end, this new class of mutations turned out to be a useful way of continuing our main interest in the involvement of genes in development by looking at the effect of nerves on early differentiation. Later, we went on to show that dominant ts lethals can be recovered and analyzed. And we continued our search for muscle defects in a new way — by screening for flies that appeared normal but couldn't — or wouldn't — fly.

There is an important lesson to be gained from the study of ts mutations. When we learn that something is inherited, most people assume then that the trait is unavoidable, there's a finality to it. Temperature sensitivity of mutations clearly shows that the environment interacts critically with genes to produce a final result. So long as a ts lethal-bearing fly is raised in the "right" environment, namely a certain temperature, it is for all intents and purposes normal.

For human beings, whose bodies and brains interact in a complex way, the psyche of a fat person in a society that prizes slimness, or that of a dark-skinned person in a fair-skinned society, will be radically affected by that physical difference. With literally hundreds of genes acting to produce our brains and bodies, clearly heredity is involved. But the environmental signals — nutrition, infectious disease, cultural experience — affect the manner in which the final person develops. It is absolute folly to search for genetic clues to individual variation if the underlying assumption is that nothing can be done to help certain groups.

The temperature-sensitive mutations did not go unnoticed; there was a great deal of interest in our results. Our work came out at an opportune time because scientists were searching for systems more complex than those of bacteria and viruses. *Drosophila,* with its highly sophisticated tools of chromosome mechanics, became attractive and once again enjoyed a golden age. Our demonstration of the existence of ts mutations added to the potential genetic tools that made fruitflies so useful. In the end, the most satisfying aspect of our work was to see ts mutations accepted as a standard scientific tool.

With outside interest in our work at UBC, the lab changed profoundly. For one thing, we started to get applications from very good students outside of Canada, and postdoctoral fellows. Postdocs are worth their weight in gold because they are fully licensed scientists. Their future careers are on the line so they are usually driven to work hard on their own. They added a large degree of intensity and professionalism, and I was relieved of many of the burdens of running the lab.

Within this diverse group, there was also a proliferation of ideas and techniques that went beyond my own expertise. I now had colleagues in the lab who were better trained and knew more than I did in some areas. I loved it! It was like being back at Oak Ridge. My grants increased quickly, while many people came with their own fellowships. As well, faculty members from other countries visited to spend their sabbatical years with us. In short, our lab became an exciting, top-notch place.

But the success of our work brought with it less desirable changes for me. With the increase in numbers, there was the inevitable change in administrative needs. I found that more and more of my time was spent writing up grant requests, filling out progress reports, and writing papers for publication — for me, the least enjoyable parts of science. The lab had once been a tight little group, and now there were factions. I found myself mediating jealousies and misunderstandings. And I was doing a lot more travelling to talk about our work. My own involvement with research at the bench shrank drastically. I soon found ideas bandied about that were simply beyond my field of specialization so I had less to contribute. I loved having the breadth and expertise in the lab, but it was no longer the personal hands-on experience for me that it had been in the early days.

In the early days, before I could afford even a technician, I used to set up massive experiments and count flies until I couldn't see straight. We had a cot in the lab so I would then lie down and sleep for a couple of hours, get up and count some more. I would do this until the bulk of the experiment was done, sometimes days later. For years after those days of my marathon three- and four-day experiments, I looked back nostalgically and dreamed of doing things that way again. (In the late seventies, I decided to try once more and set up a large experiment. But I found I didn't have the stamina. More likely than a decline in physical endurance, I had simply lost the heart for it. There was no longer the same need to prove anything, and too many other things seemed more important.)

PERSONAL LIFE AT UBC

When I first arrived in Vancouver to teach, I worked hard to establish my own group of a small band of students interested in experiencing the excitement of research and willing to spend long hours at it. During the sixties, the lab consisted primarily of undergraduate honours students and M.Sc. candidates. Using my Oak Ridge experience as a model, we held a twice-weekly "journal club" where one of us would report on a recent paper that had appeared in the library. Progress reports on current experiments and results were offered for criticism and suggestions. We also spent a lot of time drinking beer and socializing together. It was fun, and it was an enormous commitment of time.

People in the lab understood and appreciated the intricate details of experimental technique. They puzzled over inexplicable results, helped to design new tests and equipment, and revelled in successful experiments. In short, I had much in common with my students, though I was out of sync with the faculty who were preoccupied with families, buying homes and university politics. I was much more interested in science and in the students who shared my enthusiasm. I chose to ignore the faculty, to teach as well as I could, and to spend the rest of the time hidden away in the lab with my own people.

Lab life totally dominated my family life. We lived right on campus in faculty housing, so I could zip back and forth to the lab in just minutes. I immersed myself in research and teaching with a vengeance. After eating dinner and putting the children to bed, I'd go back to the lab. I seldom left the lab before one in the morning and was there almost every weekend. My extracurricular activities combined lab and family; I even took my lab colleagues along with the children on weekend outings.

I wanted desperately to be a success, to do work that would

be recognized by my American colleagues. It wasn't enough to have articles published in Canadian journals; I wanted to get papers into the most prestigious international publications. It simply was not possible for me to do my best research on a nine-to-five basis, five days a week, while teaching and doing administrative work as well.

Today, having done the science, seen the benefits but also the cost inflicted on others, it's easy to look back and think it wasn't worth it. But I doubt that my tunnel-visioned commitment would be much different if I were to have another try at it. I was obsessed, not with the idea of making a great discovery that would win fame and fortune, but with a drive to prove to my colleagues that I was worthy of their respect as a scientist. Basically, it was an extension of that childish desire to have my father pat me on the head and tell me, "That was well done." Only now I wanted my scientific heroes to admire my clever experiments.

After a year at UBC, I was spending less time at home with my family than I had while in graduate school or at Oak Ridge. It simply wasn't fair to Joane and she finally suggested that I had to decide what my priorities were. Her life was circumscribed by my demands and the day-to-day needs of two young children. I can remember sitting at the table after dinner while Joane recounted something that had happened that day. Right in the middle of her story, I suddenly blurted out, "If I make the right cross next week, then I can . . ." We stared at each other in horror as we each realized that instead of listening to her, I was thinking about genetic crosses. We were living in two very different worlds.

In the summer of 1965, shortly after our third child, Laura, was born prematurely, Joane and the children moved into our newly purchased house, while I moved into a basement apartment nearby. Laura had been conceived after we knew our marriage was over, but Joane wanted to have the baby. Our marriage had ended, not with rancour but with great sadness. Although we didn't go through the formality of a divorce until two years later, the physical separation signalled our final break.

Moving out hardly made any difference to my day-to-day life. I still worked the same hours at the lab and would drop over to Joane's house around dinner time. It was over a year before the children even realized that I wasn't "at the lab", that I had actually moved out.

What little time I did spend outside the lab was devoted almost completely to the children. Along with research, they were my priority and I spent time with them every day as faithfully as I could. I am grateful that Joane never denied me access and never used the children as a means of getting back at me. We had been sweethearts in high school, had been married for seven years and had three children together. It would have been a denial of all that sharing if our relationship had ended in hatred.

We didn't go to a lawyer to work out the details of our separation. I simply gave Joane everything we had — the car, a new house which I bought but never moved into, the furniture and most of my salary. Joane wanted to be at home with the children until they were all in school. That seemed reasonable to me. She trusted me to do everything I could to make sure she had what she needed to provide for them. When we talked to a lawyer about a divorce, he was astounded that neither of us wanted to specify the exact details of our financial arrangement. I told him Joane would have to trust me to do the best by her and the children — and she did.

Only now that I have a second family with small children, do I realize the enormity of what Joane did in raising those children alone. I thought I was a dutiful father, putting my time in with the children. But children's needs cannot be accommodated to a parent's schedule. No one can predict when children will fall down, hurt themselves, want to share a discovery or need a parent for those thousand-and-one other moments. I know the slogan "it's *quality* time, not *quantity* time that matters," and I think it's baloney. I failed with my first family and my failure was not being there in those moments when noses needed wiping, tears brushed away, discipline enforced and celebrations shared. I was too selfish and self-centred, and the children paid the price.

GRANTS AND SCIENCE IN CANADA

After my marriage ended, I had a relationship with a woman who was also a scientist. She was a very capable researcher, and because she was in microbiology, she was able to apply for grants from the Medical Research Council. Medical Research Council budgets have always been much larger than those of the National Research Council, and the MRC had a policy of giving good-sized grants to the best people. When the size of her grant surpassed mine, I went into a terrible depression, fuelled by jealousy and competitiveness. It was entirely based on gender. I didn't mind that a lot of male scientists had much larger grants than I, but it was different with my *girlfriend*! It exposed the competitive male side of me. Characteristically, *she* didn't see science as the competitive cutthroat business I saw. She took great pains to reassure me that her grant size didn't diminish my stature in her eyes.

Years later, I continued to cling to my grant, long after I had given up active research. I even applied as a co-grantee with Tom Grigliatti who was now a professor and running the lab. I finally realized that all of my social conditioning as a graduate student had been based on a highly competitive male model. Doing research is a macho business! And the most tangible evidence of one's manhood is the *grant*. The bigger the grant, the more endowed one is. The loss of the grant (or a girlfriend's bigger grant) represents an emasculation that is hard to take. Too often, aging scientists who have long passed their creative stage and have risen in the ranks of administration continue to fight to keep a grant. I hope that the increasing numbers of women in science will soon bring a radical change in this expression of machismo. We need an alternative way of satisfying the ego needs of aging scientists.

Canada has produced some remarkable scientists in spite of grant support that has always lagged behind that of the United States. Sadly, many of these scientists are no longer in Canada. Indeed, if you look at Canada's presence in the area of biotechnology, for example, we are a long way down the list worldwide. Yet, if we were able to bring all Canadians in this area back to this country, we'd be among the top five countries in the world.

What has gone wrong? I think the explanation lies in the history of the country. To achieve a worldwide audience, our most talented people in all areas have had to go elsewhere, mainly to Britain and the United States. As a consequence of this drain, those who remained behind developed a colonial attitude that anything in this country can't be as good as it is elsewhere. And we have come to feel that if one is first-rate, one wouldn't be here. We lack a spiritual faith in our own ability.

Because grants have always been small, Canadian scientists have had to scrounge materials and devise clever ways to do things on the cheap. We have never been in a position to compete head-to-head with a big American lab because of the disparity in funding, so we have traditionally developed strengths in areas that are not as high-profile. Thus, for example, Canadians have been first-rate in fisheries research, and some of the leading scholars in ecology are Canadians. We should be aware of these historical facts as we try to jump in on the current "hot" areas of research.

If there is anything positive to be said about the underfunding of scientists in Canada, it is that the administrative requirements surrounding grants are minimal. In the United States, one can spend months writing a very long grant application and having it assessed. In Canada, grant applications are simple and straightforward, and once the money is committed the recipients are free to use it as they decide. For example, when we decided to look for paralytic mutations, we were able to take a chance and try the experiment without first having to go through a complex process of getting permission.

When I returned to Canada and applied for a National Research Council of Canada grant, standard grants were 5 to 10 percent of what my American peers were getting. To be completely fair, American recipients lose between 25 to 50 percent of their grants to university administration as overhead. They can also pay themselves up to two-ninths of their regular salaries from the grant during the summer. We don't do any of that in Canada.

The Canadian tradition has been that a scientist ages into respectability. Older faculty accumulate increases over the years and end up with the biggest grants. During the sixties, two people, Lou Siminovitch (University of Toronto) and Gordon Dixon (now University of Calgary), who were both in medical faculties, waged a battle against this tradition. They fought to have the best young faculty given grants that would allow them to be competitive with their American counterparts. I believe they were very effective in helping young scientists. Certainly, my grant increased dramatically during their stints on the committee. But the NRC was still faced with a dilemma that to this day has never been fully resolved. If one big grant was given to the highest qualified applicant at the level equivalent to what he or she would get in the United States, that grant alone would wipe out ten or twelve small ones.

In Canada, we are influenced by the desire to spread the limited pot around. While I was on the grant committee, I was always infuriated when, in the name of fairness, a good scientist at a small university would be given money over an equally deserving person at one of the big universities. I believe the latter person, being surrounded by supportive colleagues, students and visitors, would on average make a much greater contribution. We seem to feel that if we spread manure over a broad patch of land, flowers will sprout up everywhere. But I believe that when we have a small amount of fertilizer, we ought to put it where the seeds are. These days there is some talk of "centres of excellence" and the need for special grants for the outstanding people. It's about time!

I had returned to Canada for reasons that had nothing to do with science. I wanted to leave behind the blatant bigotry and the urban violence of a troubled U.S. society. I vowed when I left that, much as I admired a lot of American traits, I would not return to live there. Ironically, what enabled me to stick to that vow without wavering when attractive offers from U.S. universities were made was American grant support. Let me explain.

When I first arrived at UBC, there were few of the perks that I had taken for granted at Alberta. My NRC grant was a pittance and UBC has always been squeezed for money. Within the first few years at UBC, my expenditures far outran my actual grants and I was pressured by the administration to stop my cost overruns. Yet I felt strongly that I needed equipment and personnel to do what I was doing at a reasonable level.

What saved my scientific life at that time was the year I had spent at Oak Ridge. George Stapleton, a scientist at Oak Ridge, liked me and my work. He encouraged me to apply to the U.S. Atomic Energy Commission for a grant. He was on the granting panel that considered my application and at the time when I was getting a few thousand dollars from the NRC, the American agency gave me over $15,000! It was still modest by U.S. standards, but it made me a king among my Canadian peers. I held it for four years, by which time my NRC grant had increased to a sufficient level for me to go on. It really was American support that got me off the ground.

In 1965, I received an invitation to give a lecture at the University of Hawaii. There were some top-notch people in the Department of Genetics in Hawaii, and the islands were wonderful. The head of the department told me to rent a car and drive around Oahu on my own. The night I was to return to Vancouver, I was fêted by some of the faculty and offered a job. At the airport, the department head asked if I would accept and I said yes. Besides a substantial salary increase, the job offered an incomparable setting.

When I got home, I told Bill Hoar, the Head of Zoology at UBC, that I had accepted the job at Hawaii. He immediately asked whether I had done so in writing. When I said no, he asked me not to do anything until he had arranged for me to see the president of the university. I went to meet President John B. MacDonald, now head of the Addiction Research Foundation in

Toronto. Remember, I was a green assistant professor and this was all pretty heady stuff. MacDonald asked what the attraction was at Hawaii. I told him it was *not* the salary increase at Hawaii, but rather the money for research. MacDonald then figuratively reached under the table into a discretionary fund and offered me $10,000 towards my research costs. That sealed it; I turned down the Hawaii offer. It was the closest I ever came to leaving Canada.

There was one other time I was tempted to return to the United States. In 1970, I gave a seminar at the Massachusetts Institute of Technology in Cambridge. MIT's Biology Department members have the attitude that if you are good enough to be hired there as an assistant professor, you are tops, and the department does everything it can to make sure you do well and gain tenure. So cooperation and goodwill radiate within the faculty.

I loved the excitement and easy communication that pervaded biology at MIT. My seminar there went well and the head of the department asked me whether I'd be interested in a tenured, full professorship. I was very flattered and excited and said, "Sure." The next day, I visited Jon King, a bright young assistant professor who had graduated from Cal Tech in viral genetics. He said, "I hear you're interested in coming here." When I affirmed this, he surprised me by asking, "Why?" I stammered out that it seemed obvious. It's one thing to make a splash in the boondocks like Canada, but I had always wanted to see if I could make it in the big time. His reply changed my attitude and profoundly affected my life. He said, "You don't have to come *here* to find that out. You've *already* made it. That's the reason they want you here." He went on, "People at MIT are so good — students, the faculty, visiting scientists — that anyone would make it here. The good people around will carry you. If you've made it in the boondocks, you can make it anywhere." I've always been grateful to Jon for that, and never again seriously considered leaving Canada.

SCIENCE AND POLITICS

High technology these days beckons to politicians as the new panacea — holding out the promise of carving a place for Canada in the international market. It's remarkable how, after supporting the scientific community on a starvation diet at least for as long as I've been back in Canada, our leaders now trumpet forth about our exceptional technologists. To be sure, there are people here who are as good as anyone anywhere else in the world, but they have managed under enormous disadvantages. They are far too few to make claims that we can cut big inroads into markets dominated by Japan, the United States and Europe.

Politicians must, of necessity, limit their vision to the interval between elections. So when they see a project holding out promise of economic benefits, they are willing to put money into it. But they must see a return before the next election. This cannot work. Groups of creative scientists cannot be quickly assembled to automatically generate results with production-line regularity. Scientists have to feel the confidence that they are appreciated and can rely on long-term grant support that will allow them to compete with colleagues in other parts of the world.

Too often, Canadian scientists set limits on their outlook. It gets my hackles up when I hear, as I have in my department, that "we are the best Zoology Department in *Canada.*" I'm interested in comparing our department and our people with the best in the *world.* Too often, Canadian scientists deliberately cut down the amount they request for research because they know that there isn't enough to get what they really need. They don't fight for what they need; they make do with what's available. That's not how we develop an eminent community of scientists. It will take a long time to create a climate within which top minds can flourish.

Politicians simply don't understand this. Before he left office, Mr. Trudeau presided over the establishment of the Biotechnology Research Institute with a capital grant of $51 million to build a centre in Montreal. This was supposed to inject us into the hot field of genetic engineering. But he got it backwards. First we should have identified the people, got them together into a unit, funded them well and *then* worried about buildings.

British Columbia's former minister of science, Pat McGeer, who is himself a scientist, was intrigued by the high tech of Silicon Valley. Mr. McGeer wanted to have a Silicon Valley in British Columbia too. His plan? He set up multi-million dollar Discovery Parks — land on which new high-tech companies can be built — next to UBC, Simon Fraser University, the University of Victoria and the B.C. Institute of Technology. So now Silicon Valleys are supposed to sprout up in each park. He should have known better. The reason Silicon Valley exists is the presence of top people at Stanford University and the University of California at Berkeley.

At both the federal and provincial levels, high-tech is the buzzword, but there is no evidence that it is more than a shortsighted fad. In Ottawa, the power of members of the Cabinet is measured by the size of the budget of a minister's portfolio. In 1971, Prime Minister Trudeau established a new portfolio, the Ministry of State for Science and Technology, or MOSST, with Allistair Gillespie as its first minister. By the criterion of budget, it is a minor portfolio and many ministers hold it in addition to a major ministry. A succession of ministers in MOSST seldom held the post for more than a year. The department was to provide an overview of science and technology in all the other departments, but in reality it has very little clout.

In 1979, during the shortlived Clark government, Heward Grafftey was appointed head of MOSST. I had met with every one of the ministers, either as a broadcaster or consulted scientist. With the exception of Hugh Faulkner, I had not been impressed by them or hopeful that they would be able to help scientists. But when Grafftey came in, there was a real sense of excitement and change. Within weeks of his appointment, Grafftey asked me to come to Ottawa to share ideas about science. I spent a couple of hours talking about the need to support excellence and to increase grant sizes to a level that would allow our top scientists to be competitive with scientists in other countries. He listened carefully and was very supportive of my ideas and enthusiastic about increasing government commitment to science.

I was surprised a few months later when Grafftey called to tell me he wanted *me* to take a position in government so that I could actually act on some of the things we had discussed. William Schneider, the president of the NRC, was approaching retirement and Grafftey suggested that I could have the job. Now I am not a wheeler dealer — if I have any ability, it is to be provocative and to stimulate people to think about things from a different perspective. I thought I would make a terrible administrator and bureaucrat, so I immediately thanked him for the honour and turned it down. Besides, I told him, while I felt it was time to see the Liberals out of office, I was not (said vehemently) a Tory. Moreover, I would not give up my job as a television host which I felt was the most important activity I did.

In a few days, Grafftey was back on the phone again. Prime Minister Joe Clark was willing to back him on whomever he chose, Grafftey told me, and Grafftey wanted to submit my name for a position. He even felt we could work out a way for me to remain on television. Then he informed me that Gordon McNabb, the Liberal-appointed president of the NSERCC (Natural Sciences and Engineering Research Council of Canada) which now administered all grants for universities and the NRC, was to be shifted out of that post. He wanted me to consider the job. I immediately turned it down, but he persisted and pointed out that if I wasn't willing to put my actions where my mouth was, I had no right to spout off in public. Here was a chance to really do something — to get funding for centres of excellence and to try to build pockets of top people. It was an agonizing dilemma.

Finally, I called a trusted friend, Gordin Kaplan (now vice-president of Research at the University of Alberta) and he advised me to go for it. I talked it over with Jim Murray, executive producer of "The Nature of Things," and he thought it would be pretty unlikely that I could stay on as host if I took the job. I talked it over with Geoff Scudder, the head of the Department of Zoology at UBC, and he agreed to get me a leave of absence from the university.

After a lot of soul-searching, I finally decided it had to be tried. So, on December 13, I called Grafftey and told him I would let him put my name forward for the job. That very night, the Clark government fell and I never heard from Grafftey again about that matter. In the election that ensued, Grafftey lost his long-held seat in Montreal; I watched him being interviewed on election night as he bitterly accepted the judgement of the people. I believe he might have done a lot as the minister of MOSST.

When Mulroney was elected prime minister, his new head of MOSST, Tom Siddon, invited me to Ottawa for a discussion. I was delighted to meet Grafftey again and he was as exuberant as ever. There is life after political office! Fortunately for the scientific community, Gordon McNabb held his job at the NSERCC and was a courageous and effective fighter for his constituency. I never for a minute regretted losing that opportunity.

Science delivered thrills, honours and rewards beyond anything I ever dreamed while in graduate school. Few people in society have the opportunity or the freedom to spend a period of time totally consumed in an activity, but there is a dark side to this way of life. Other relationships — friends, mates, children — must accommodate the overriding priority of the lab. (Of course, this is not unique to science as politicians, high-powered business people and lawyers know.) As well, the explosive growth in the scientific community, accompanied by more information, techniques and competition, requires greater concentration within an increasingly restricted field. In a time when more than ever before the application of science is having enormous consequences for everyone, scientists themselves are more myopic than ever.

Dad's father and his great-grandchildren (l to r): Laura, Troy and Tamiko.

Salmon at a hatchery in Washington State, used to collect salmon pituitary glands. (November 1964)

Twelfth International Congress of Genetics in Tokyo, Japan, August 1968. (l to r): Arnold Ravin (U.S.A.), Gordin Kaplan (Canada), me, unidentified (Italy), Alan Campbell (U.S.A.)

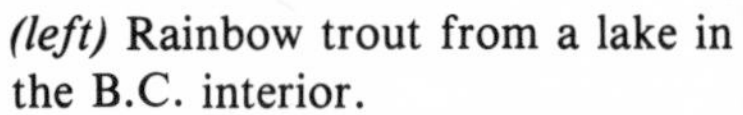

(left) Rainbow trout from a lake in the B.C. interior.

(right) Troy and trout caught in Garibaldi Lake, B.C. We had to backpack seven miles up the mountains.

A growing Suzuki clan at Christmas, 1971: (From l to r): *Back row*: Alex Szlavnics (Aiko's husband); Aiko; Dawn; me; Dick Aoki; (Marcia's husband); Tamiko. *Middle row:* Dad in a gag pose with a trout; Troy; Mom; Cica on Mom's lap (Aiko's daughter); Marcia; Joane. *Front row:* Laura; Jill and Janice (Marcia's girls).

(left) The lab in the late '60s was dominated by undergraduates and M.Sc. students. (From l to r): *Front row:* Tamiko, Ruth Dworkin, Udo Erasmus, Mike Schewe, Burt Ayles, Kathy Yang, Laura, Barbara Suzuki. *Middle row:* Troy, Mary Tarasoff, Jill Cameron, Doug Procunier, Don Guichon, Rachel Pratt, Shizu Hayashi, Diana Combe, Jeanette Holden, Leonie Piternick.
Back row: Rod Williamson, Lorne Shaw, me, Ella Korinek, Earl Rubin, Lyn Baldwin.

The children (l to r): Laura (age seven), Tamiko (age twelve), and Troy (age ten) and their beloved cat, Tomasina.

(left) The lab at the peak of our activity in the fall of 1973. (l to r):
Back row: Don Sinclair, Malcolm Fitz-Earle, Jim Stone, Gordon Wong, Jim Taylor (Marla's husband), Ian Duncan, Daryl Falk, Shirley Macaulay, Dad, John Macaulay, David Sheppard, Tom Grigliatti.
Middle row: Bruno Jarry, Mrs. Jarry, Tom Kaufman, Ella Korinek, Tara Cullis, Oksana Suchowersky, Nicole Follea, Paul Bauman.
Front row: Len Kelly, Raja Rosenbluth, Mrs. Sheppard, Marla Kiess, Elaine Tasaka, Mom, Mrs. Fitz-Earle, Jan Grigliatti.

CHAPTER SEVEN

TEACHING — MY FIRST AUDIENCE

A UNIVERSITY is a special and unique institution. It is a community of people who are drawn together by their belief in ideas and the power of the human mind. A dynamic university supports a bewildering variety of people, some outside the norms of social convention. Scholars, dreamers, inventors alike inhabit a university. Exchanges of ideas, open discussion and tolerance are the cornerstones of university life.

But often a university falls short of these expectations because it is, before all else, a collection of human beings. My experience of university life has run the full range — from the highs of shared discovery and insight, to the depths of deceit and hypocrisy. But it has always remained a haven, a base from which I could carry out my other activities.

Ever since my last year at Amherst, I knew I wanted to teach. When I first started graduate school, I didn't think I was good enough to be a "real" scientist, but I knew I could be a good teacher. As I gained confidence as a scientist, research became my main interest, but teaching remained an important part of my life. Teaching kept me in touch with students who, in turn, led me to explore ideas and questions which I might otherwise never have thought about. I also suspect that my lecture "performances" in class were a good training ground for my eventual career in television.

When I arrived at the University of Alberta as a brand-new assistant professor, I thought teaching would be a snap. I had had some experience in teaching as a lab assistant in courses at the University of Chicago. As well, I had given a number of scientific lectures at meetings. Yet when I started to prepare lecture notes for the first time, I discovered I didn't really "know" the material, even though I was a fully licensed geneticist with several scientific publications on my record.

I tried to put myself in a beginner's place and to develop a sequence of lectures that would peel away the layers of mystery, just as Professor Hexter had done for me. In the process, I discovered nuances and subtleties in genetics that I had not appreciated before. As long as I was teaching, I learned new things. Each year someone would ask a question that I never had occasion to answer before, or else approach a subject from a completely new angle. That is what keeps teaching fresh and challenging.

The first course I taught was for second-year students working on a degree in Agriculture. Initially, I felt it was demeaning for a hotshot like me to be teaching mere "Aggies." But that class turned out to be the most memorable one I ever had. The students were bright and keen and did everything with gusto. I demanded a lot from them and they responded with everything they had. Most of them had grown up on farms and were accustomed to hard work, so they put out for me.

Teaching was far more to me than just walking into class, giving lectures and posting office hours — it was a way of life. During this time, I became very involved with the lives of my students. I spent hours listening to them, as they discovered a whole new world of ideas, cried over broken love affairs and tried to find long-term goals. Often my students would drop by the lab at night and drag me off to a local pub. Almost every weekend there would be Aggie parties, and my wife and I would be invited. They regaled me with stories about farming, and in spite of my disdain for the practical side of genetics, they got me interested. As far back as 1962, we speculated on the feasibility of cloning cattle and genetic engineering in wheat. Those students became my good friends, and for years after they would write to me in British Columbia, send cards and even come to visit.

The Aggies pushed me, against my will, to speculate on the value of animal and plant breeding. Genetics was interesting to Aggies, because it affected their futures in farming. On the other hand, I wanted to show them genetics as an intellectually exciting activity. Applied genetics was completely uninteresting to me and seemed rather vulgar, but the students' interest forced me to open my mind to this side of genetics.

I got my feet wet in teaching that first course in Edmonton. Those Aggies were my guinea pigs — I worked out the bugs in the course on them and I'm sure I learned far more from it than they did. In return, I gave them all the energy and enthusiasm I had. But even before that course was over, I had decided to move to UBC. I doubt that any subsequent class could have lived up to that first one in Edmonton, so I'm glad it has remained unique. And when I arrived in Vancouver, I was a much better teacher because of those budding farmers and ranchers.

TEACHING AT UBC

At UBC, most of my students were prospective zoologists or pre-meds. The former group was often interested in population genetics and evolution, while the latter was fascinated by new insights into human genetics. Again, these were areas towards which I, as a chromosome mechanic, had always been rather contemptuous. But I had to investigate them in order to answer the students' questions. It was the students, once again, who pulled me out of the rarified heights of fundamental genetics into some very crude but far-reaching techniques of applied genetics.

I began to learn about methods of prenatal diagnosis. Although amniocentesis had first been used in the forties, it was still in its infancy and we never dreamt it would become the powerful and routine technique it now is. I knew about cloning because it had been done in spectacular fashion in the fifties using frogs. While mammalian eggs were much smaller and therefore harder to manipulate, it was clear that egg size was just a technical detail that would be overcome. Gradually, I came to realize that the potential application of genetics for deliberately engineering farm animals, plants and people was more than just the stuff of grant proposals or science fiction — it was a very real possibility. As I discussed the implications of genetic engineering and cloning in my lectures, the students responded with enthusiasm, often posing challenging questions. Their initial interest started me thinking and their responses continued to stimulate my elaboration of the ideas.

When I arrived at UBC, genetics was a second-class subject. The Department of Zoology had made its reputation in classical subjects, such as ecology, wildlife biology, fisheries and physiology, while genetics was taught in the Botany Department. Most of UBC's zoologists were interested in "whole" organisms, not the modern cellular and genetic fields; however, they eventually recognized the importance of genetics and I was hired as a faculty member of their department.

Zoology was a much larger department than botany — it offered more courses and had many more students — and there had been rivalry between the two departments for a long time. Before my arrival, genetics was taught by a woman in botany who had been classically trained in cytology, the study of cells and cell structure. She was a recognized scientist in her field, but her course was rooted in the classical genetics she had learned before the microbial and molecular genetics of the fifties and sixties. My course quickly became part of the power struggle between departments. I represented the new way, and was a threat to her course, as well as to her department.

Initially, the Zoology Department pushed through my new genetics course under the guise of its importance to Zoology honours and graduate students who needed special emphasis on heredity in animals. In fact, however, I paid no attention to the arbitrary lines between animals and plants in my teaching. I believed, and still do, that genetics applies across these biological categories.

My course was up-to-date and demanding, and my twenty-five honours and graduate students appeared to enjoy it. The next year, word of the course spread and more than eighty students signed up for it. Many were not eligible for my course but wanted to take it instead of Botany's version. So I pushed to have my course opened to any student who wanted to register. With all of the conceit of a young faculty member, I felt my course was superior to the "old-fashioned" approach in Botany's genetics course and thought students who didn't take mine were being cheated of all the modern stuff.

The botanists fought to preserve their turf by keeping my course restricted to Zoology honours and graduate students. At one point, I met the head of the Botany Department, who said to me, "David, you're young and ambitious and you want things done right away. Take it easy. Wait ten or twenty years and it won't seem so important." Now, twenty years later, it's hard to remember why it all seemed so urgent. Most of the students taking genetics were not prospective geneticists and what mattered was that they got a taste of the subject. The ones going on in genetics would quickly get what they needed on their own or in advanced courses. But at the time, this battle was very important to me, and within a year my class was open to all students. Happily, for several years now, genetics has been taught at UBC as a joint effort by professors in both departments.

When I arrived at UBC in the fall of 1963, the Zoology Department was still a very formal place. My informality set me apart from the rest of the department right from the beginning. Faculty members came to work in jackets and ties, and their students addressed them as "Doctor." That was far different from my experiences in the American schools I had attended. I didn't wear a jacket and tie, and encouraged my students to call me Dave, much to the disapproval of the faculty. Blue jeans were my preferred style. I suppose some people may have felt I was pandering to students by trying to be one of them, but the fact is I had worn jeans since high school.

The elitism of the British class system was carried over into the University of British Columbia. There was, and remains, a strong sense of stratification in the university, with administrators on the top and students buried at the bottom. I had been shaped by the egalitarianism of the American universities. In 1963, I was an oddity who didn't fit the mould of a UBC faculty member.

Just as I had been an outsider among the *Nisei* kids in Slocan, as a brain and social misfit in London, and as a poor person in the rich man's school at Amherst, I was once again an outsider as a faculty member at UBC. But now it was different. I believed that a university was a place of tolerance, a place where respect could be earned by ability and hard work. I really didn't care about fitting in. Now I was an "outie" by choice. My students gave me all the enthusiasm and admiration I needed, and they were far more tolerant than my fellow professors. To be honest, I was also pretty arrogant — I believed I was as good as anyone in the department, that I could play by my own rules. I was interested in establishing a reputation internationally, not in the puddle of department politics.

It was one thing to teach a class of undergraduates, and an entirely different matter to deal with them in the more intimate conditions of my lab. When my students came in, I tried to foster the same exuberance and intimacy that I imagined Curt Stern had at Berkeley. My students and I shared our enthusiasm for genetics and a willingness to put in time working. While I was clearly the head of the lab, we were a tight group.

However, I never acquired the knack of being close friends with students while still maintaining the necessary delineation between teacher and student. So when I had to enforce a deadline for an experiment, an essay or draft of a thesis, it was often a shock to the student. Suddenly, Dave the buddy became Dave the heavy. For the most part, students were mature enough to accept this reality. But in a few cases, there was a sense of betrayal. Closeness had only masked the true hierarchy of teacher and student.

At the beginning of my teaching career, when my classes were small, I was able to impose the kind of expectations that were laid on us at Amherst. And the students rose to it; they performed well. The best of my students at UBC were as good as any in the world. It delights me to meet some of those former students who now work in universities, government labs and private companies.

But once my course was opened up to anyone who wanted to take it, enrolment gradually rose into the hundreds. My course then had to change. With large numbers, all hope of intimacy with students was gone. I tried throwing beer parties at the Faculty Club, but you can't get to know people when there are more than ten or twenty in a room. A the end of the course, students would appear whom I didn't even recognize! Where I had once known every student by name, I now saw anonymous numbers. Teaching was not the joy it had been in the first years of my career, and the students suffered as a result. My very success as a teacher attracted more students and this caused the loss of what had made my course successful — the sense of intimacy.

The part I regretted most was that there was a qualitative change in the atmosphere of the class. A university course is more than just the transmission and receipt of information. It embodies the teacher's personality; it contains the distillate of his scholarship, research and introspection. Each lecture is a sharing of ideas, a very personal gift from the teacher. In large lecture halls, the intimacy is gone; my classes became performances rather than a sharing of ideas.

SINGLE AGAIN

In 1965, when my marriage broke up, I was single again for the first time in over seven years. Having a mate had become a powerful habit, and I felt I needed someone. Each date became a new potential mate because I wanted desperately to have a "permanent" relationship again. In fact, this wouldn't happen for seven years, in large part because it took time for me to figure myself out. But I was looking the whole time.

The seven years of my second bachelorhood were turbulent and full of upheavals. A great deal had changed since I was a teenager. The pill, the Beatles and drugs were the most obvious signs of a tremendous shift in social and sexual morés. With the exception of a few arranged dates in college, I had only gone out with Asian girls and still suffered from a self-consciousness about being Japanese. So it was with great trepidation that I again entered the ranks of bachelorhood.

The sexual revolution also meant that I had some other adjustments to make. Women expected to be treated as equals. In my family, a male signalled an empty rice bowl by holding it in the air in order to have a woman — my sisters or mother — fill it with more rice. When I once did that without even thinking at a date's home, she recoiled and demanded, "What the hell do you want?" I realized that my attitudes towards women would have to change.

The sexism I had absorbed from my parents' cultural background caused enormous problems during this stage of my life. The women who attracted me were simply not willing to put up with my chauvinism. I could understand the demands for equality in pay, opportunity and rights, because the plight of women was so clearly comparable to that of minority groups that I had supported. But the difficult part is the changing of long-practiced personal habits. Who cooks, cleans up, drives the car, keeps the accounts, scrubs the floor, and so on? Today, I would never call myself a liberated man; it will take more than one generation of awareness to excise the assumptions and values that have accumulated over centuries. But compared to where I was in the mid-sixties, I've come a long way, baby!

My eventual insights into my sexism and insensitivity came at great cost to others. I once lived with a woman who was an avid outdoors person. One weekend we had planned to go on a hike with my children up one of the local mountains. On the day of the trip, one of the kids got sick while the other two decided they wanted to do something with friends. I was disappointed but then brightened at the unexpected bonus of time. I went straight to the lab where I worked all day. My girlfriend was devastated because to her the trip had been an occasion to do something with me, while to me it was a chance to be with the children. It was simply the same thing I had always done — exactly what I wanted.

The women I dated during those years still had to fit their lives to my schedule, which continued to revolve around the lab and my children. In 1969, after the break-up of one long-term relationship, I decided to leave Vancouver for a few months to recover from my personal turmoil. I was soon embroiled in the social upheavals taking place in Berkeley.

CALIFORNIA

The sixties and early seventies were a time of optimism; students assumed a lot of power. It had been student activism over Vietnam that eventually forced Americans to face up to their role in an immoral war. "Flower Power" swept the continent, as young people questioned society's values.

In 1967, the Genetics Society of America held its annual meeting at Stanford University in Palo Alto, California. These meetings are marvellous opportunities for students to meet professional geneticists and to hear some of the leading scientists in the world. So I always tried to take as many students from the lab as possible to the meetings. I rented two cars and we drove in tandem. It was an exciting time to arrive in California. The hippie movement, with Haight-Ashbury in San Francisco at its centre, was in full bloom.

At the meeting, someone mentioned that there was a terrific place called the Fillmore Auditorium in San Francisco. So one night, everyone from the lab drove into the city to check it out. The group playing was an up-and-coming band from England, called "Cream." Their music was so loud that the whole building seemed to be pulsating as we walked in. Inside, movies (including Norman McLaren's famous National Film Board short, "Neighbours") were projected on the walls along with patterns created by mixtures of oil and coloured water squeezed between watch glasses. The air was heavy with the smell of marijuana, and naked people were being painted in psychedelic patterns with paint that glowed under the ultraviolet lights. This was the legacy of Timothy Leary and totally new to us. We returned to Vancouver profoundly affected by the movement that would sweep the continent. That glimpse of what was happening in San Francisco was a hint of another stage to come in my life.

In the spring quarter of 1969, I was invited to teach a genetics course at the University of Calfornia at Berkeley, with the possibility of a more permanent position. It was still a tumultuous time. The university was a hotbed of anti-Vietnam and radical activity, the symbol of the wave of student revolt that was shaking the power structures of North America. Ronald Reagan was then the governor of California and he was determined to bring the University of California to heel. The campus at Berkeley was electric. Students lounged around the fountain in Sproule Plaza, smoking dope or tripping out on acid, while speakers would harangue the assemblage on important issues of the day. Telegraph Avenue was lined with shops peddling all of the accoutrements of the freak movement — bell-bottoms, beads, hash pipes. As a visiting faculty member whose only responsibility was to teach a genetics course, I had plenty of time to soak it up. I loved to walk down that street each day after class just to look, listen and smell what was going on.

Just for the hell of it, I stopped shaving. And, blow me down, there were enough hairs on my lip and chin to be visible after a couple of months. I had left Vancouver clean-shaven, with short hair and dark horn-rimmed glasses. I later returned to UBC wearing a headband, long hair, granny glasses and a tiny moustache and beard. I looked like a hippie. My fellow faculty members at UBC were not amused.

A few blocks down Telegraph Avenue was an empty lot where people used to park their cars. Gradually, "street people" began to inhabit the lot. They planted flowers and camped there overnight. It soon came to be called "People's Park" and was perceived by authorities as an attack on the notion of private property. One night, policemen tossed out the people sleeping in People's Park and threw up a chain fence around it; private ownership had been reestablished. The next day, a huge crowd of angry students assembled at Sproule Plaza at noon. I ambled over to watch what would happen. A number of people got up to excite the crowd with speeches about the political significance of the park. The last speaker was the student president who worked the crowd into a frenzy and ended up with, "This calls for action, not talk. That park belongs to us. Let's go and take it!" And with that, the entire mass of people moved as if they were a single organism.

I was in the middle of the crowd and was swept along as an observer. It was exhilarating and scary. I am sure there were other faculty in that crowd along with graduate students and street people. We were all swept in a surging wave. We arrived at People's Park to find it surrounded by armed guards. The crowd split up into uncoordinated groups along different streets. The students taunted the guards, who eventually fired rounds of tear gas into the air and forced the crowd to run away. It got to be a game: we pushed towards the fence, the guards fired gas canisters, we ran away. Then we would turn around and go back. Finally at one point as we were running, I felt a stinging in my legs and realized that the police were firing straight at us with buckshot! That day one student was blinded, another killed, by shotgun blasts.

That was it for me! I'd had a look at a genuine confrontation and wanted to get out of there. People opened their houses along the streets so that anyone could run in for shelter. I dashed into a dorm and heard a guy yelling into the phone, "I got gassed! I got gassed!" as if he had passed some rite of manhood. I went back into the street, only to encounter the final indignity. Jeeps came by with a strange-looking pipe hanging out the back. It was dispensing "pepper fog," a gas which burned like pepper in the lungs when inhaled.

The campus was in an uproar. That weekend, students in Sproule Plaza were gassed from a helicopter. It was horrible — Americans were attacking their own children! On the following Monday as I walked to class, tear gas clung to the ground.

Pedestrians stirred it up again and by the time I arrived, I was weeping. The students called a boycott of classes on campus, and so, like many other faculty members, I had the students come to my apartment for lectures.

The Student Council called for a protest march on Easter weekend and there were rumours that Reagan was using the crisis to force an armed confrontation. I went on the march, mainly out of anger at the way students were being treated, but with great trepidation because of the threat of violence. As we marched, you could see lookouts on the rooftops armed with machine guns. Surrounding People's Park were the dreaded police from the Alameda County Sheriff's Office. I was relieved that it was a peaceful demonstration with a festive atmosphere. Marchers threw symbolic daffodils on the fence until it turned yellow. The incongruity of troops with flowers hanging from their gun muzzles was unforgettable.

From this experience, I realized again why I had left the United States to return to Canada. Much as I loved Berkeley and the Bay area, I didn't want to live there. As soon as my course was over and grades turned in, I came home, excited by the radicalism of students, but grateful for the sanity of Canada.

STUDENT RADICALS IN CANADA

Student activism took a few years to reach Canada. Canadian students often look to their American counterparts to see what is in vogue. After the civil rights movement in the United States cranked up, Canadian students sent money and people to the southern states while ignoring the terrible plight of native Indians and Inuit within their own country. The anti-Vietnam War movement was popular in Canada, but no one protested Canadian complicity and enormous profit-making from the war.

The radical movement hit UBC in the late sixties in the form of a charismatic visitor, Jerry Rubin. Rubin was a spellbinding orator. At UBC, he whipped the students into a frenzy and then funnelled that energy into a takeover of the Faculty Club, the symbol of university elitism. He led a march across campus, then left before the students had actually occupied the building.

The UBC faculty was outraged. I remember one of the full professors in Zoology shouting that the student occupants ought to be lined up against a wall and shot. National television broadcast pictures of students lighting joints from burning dollar bills, looting liquor cabinets and swimming naked in the fountain. Classes were cancelled as faculty met with students in "teach-ins" to talk about all the things that were wrong with the university and ways we might correct them.

There has never been another time when issues such as teaching and education so dominated faculty meetings. Usually we discuss salaries, office and lab space, and teaching loads. The Faculty of Science held a meeting to hammer out an official position on the takeover of the Faculty Club. I was still a young professor, but after listening to the senior faculty, I couldn't remain mute. I stood up and suggested that perhaps the students had a legitimate case. If there really is such a thing as a *community* of scholars, I said, then the Faculty Club with special dining rooms and exclusive membership didn't belong on campus. At that point, Charles McDowell, the head of Chemistry, and one of the most powerful people on campus, rose and savaged me with an attack on the shallowness of my analysis. I can't remember a word of what he said, but I do know that it seemed to convince everyone else in the room. I shrivelled to the bottom of my seat and sneaked out as inconspicuously as possible at the end of the meeting.

I was regarded by many students as an ally for their cause. At one point, students held a panel discussion about university reform and I was asked to present a faculty position. Carey Linde was the student vice-president who had been elected on a radical slate and he sat next to me at the table. When my turn came, I charged that students were being short-changed because teaching was a low priority at the university and suggested there was a need to change faculty assessments so that teaching could be taken more seriously. I proposed ways for students to increase their power within the university.

I knew my speech would infuriate my fellow faculty members

but was pleased with myself for showing solidarity with the students. So I was astonished when a student immediately stood up and denounced me. "You've just spouted off typical liberal bullshit!" he shouted. "You're not proposing any radical change in the power structure. You're trying to co-opt us. *You're* the enemy!" I was destroyed. Carey Linde looked over at me, shook his head sympathetically and murmured, "You're too old, Dave." At thirty-three, I realized the meaning of "generation gap" for the first time.

One of the important issues the students were raising was the gulf between what we were teaching them and what we actually did. As my father had taught me, we are what we *do,* not what we *say,* and the students know that. Universities had grown rapidly in the early sixties, and research and scholarship were the main factors considered in promotion, tenure and pay increase. "Publish or perish" was not an idle slogan. So to many faculty members, teaching was a low priority, a duty that interfered with what they considered the more important activity of research. As well, many introductory-course classes were huge — lecture halls were jammed with hundreds of students. For personal contact, undergraduates often approached graduate students or teaching assistants hired on short-term contracts. I was always shocked to hear students refer to a graduate student instructor as "my prof."

Universities and departments have never resolved the fact that, though large undergraduate courses are the main justification for their operating budgets, those students are faculty's lowest priority. Universities have become huge factories where undergraduates may pass through four years without ever having an intimate conversation with a professor. There were powerful reasons for student dissatisfaction. There still are. In the sixties and seventies, students were demanding more power in shaping university priorities and in the allocation of its resources. I regret that they achieved so little in the end.

One person I came to admire greatly during this period was Malcolm McGregor, the head of the Classics Department. To the students, he seemed like someone out of the Middle Ages. At one point, he agreed to debate with the students about what a university should be. He offered an elitist, classical position that was close to what I had experienced at Amherst. I agreed with most of what he said, though I was still caught up in the then-fashionable rhetoric of letting "everyone do his own thing." The audience was openly hostile to McGregor, but he revelled in his role as bad guy and defended his position unequivocally to loud boos.

One of his fellow panel members was an undergraduate who kept referring to him as "Malcolm." McGregor kept shooting dirty looks at him and finally bellowed, "You will call me *Mr. McGregor!*" The cocky student radical was completely cowed and did as he was ordered. McGregor's courage and conviction were rare during those times.

I still believe that there was a great deal of merit in the issues raised during that brief period in the late sixties and early seventies. But when students began to tell me that the overriding consideration of everything in university had to be "fun," I tuned out. The true rewards came from discipline and hard work.

HIRING PROFESSORS

In the late sixties, North American universities were expanding at a terrific rate and we were all competing for the same pool of potential faculty members. Canadian universities simply couldn't match the offers of frontline American schools like Harvard, Stanford, Cal Tech, Yale and MIT. By virtue of their prestige, research facilities and equipment, they skimmed off the top candidates and Canadian universities were left to select from the rest.

I had left the United States to return to Canada because Canada was home to me, a place where I felt comfortable and wanted to make a contribution. The eight years I had lived in the United States had added a broader perspective within which I could see Canada more clearly. Canada mattered to me as a country that was different from the United States.

Today, as Canada attempts to compete for a part of the global economy, we must, above all, believe in our own ability. Youngsters growing up in any society need role models to look up to, to admire, to emulate — to affirm that people in their society can compete with the best. That's why it's so important to have the Northrop Fryes, Margaret Atwoods, Karen Kains and Wayne Gretzkys. Youngsters also need to know that it's possible to become a scientist and compete while remaining *in Canada.*

We have so little confidence in ourselves that we wait for confirmation of our best by other countries. When our actors and directors and writers find jobs in Hollywood, we proudly proclaim them as outstanding Canadians. We do the same in science, taking a perverse pride when our good people are offered jobs at Harvard and other top schools. Then we know they *must* be good. Too often we only realize that after we lose them.

As Canadian universities scaled up to crank out Ph.D.s in the late sixties, this lack of faith in the quality of our own continued to apply. Many of our department heads and deans were themselves from Britain or the United States, and simply took the superiority of Ph.D.s from their own countries for granted. The irony is that while the cream of Canadian talent was being lured away by the Harvards, Stanfords, Oxfords and Cambridges, those universities outside Canada were also skimming off the best from the United States and Britain. So while we were losing top Canadian scholars, the foreign Ph.D.s we were getting were not of the same academic quality. Governments were only concerned with numbers; so as long as there were the same numbers of Ph.D.s coming into Canada as were leaving, they perceived no brain drain.

In 1969, Robin Matthews and Jim Steele published their classic book, *The Struggle for Canadian Universities: A Dossier,* which looked at the steady de-Canadianization of faculty in Canadian universities. Canada is in a class by itself in the percent of non-citizens it hires for academic positions. No other country, including those in the Third World, hires permanent staff as freely from the rest of the world. Even cosmopolitan Harvard University has no more than 10 percent non-American faculty. While Canadian graduates are discriminated against in the United States on the basis of nationality, they must compete on an equal basis with American applicants in Canada. In my department when I arrived, 52 percent (eleven out of twenty-one) of the faculty were Canadians. Today, 42 percent (sixteen out of thirty-eight) are Canadian.

In the mid-seventies, I proposed at a department meeting that we adopt a new policy. I suggested that when we screen applicants for a job, we set up two categories: Canadian and Other. We would first evaluate the Canadian candidates, and if any of them met our requirements and standards, we would offer the job without even looking at the candidates on the other list. Only if we failed to find anyone qualified would we look at the other file. The head of our department, who was from Britain, let the other faculty members have their say. One full professor (an Englishman) called me a fascist. One of the associate professors, also English, was almost inarticulate in his rage. But what astonished me most was the reaction of the Canadians. It is often Canadians who have studied elsewhere who are the most prejudiced against fellow Canadians. They themselves are convinced of the inferiority of Canadian education and favour open competition regardless of nationality. I was only suggesting, in a more formal way, what is *de facto* practice in the United States and Britain. And I was still trying to ensure that by favouring Canadians, we wouldn't compromise standards. My suggestion came to naught, except to create another wall between me and some faculty members.

I know of Canadian scientists who have tried desperately to return to Canada, only to meet an indifferent or even hostile response. Bob Edgar is a classic case. He pioneered the development and use of temperature-sensitive defects as an important class of mutations in viruses. When he graduated from Cal Tech, he heard that there was an opening for a geneticist at Queen's University. Edgar applied for the job and received a reply suggesting that he was applying for the wrong position — the job was for genetics in the Biology Department, whereas the Queen's people had classified Edgar as a virologist. Edgar was dismayed; he considered himself to be a geneticist who happened to be working on bacterial viruses. So he took the professorship he was offered at Cal Tech and went on to great fame and glory as a geneticist. He eventually became the provost at the University of California at Santa Cruz. It is regrettable he was not able to return to teach in Canada.

Another example is Henry Taube, a Canadian at Stanford University who won the Nobel Prize in 1985 for his work in chemistry. When he graduated from an American university, he had wanted to return to Canada and wrote to over a dozen Canadian universities. He was turned down by all of them. He therefore took a position at Stanford where he did his great work. There are many similar stories. I take no pride in being able to point to Edgar and Taube as *former* Canadians who couldn't find a position in their own country.

TARA

For me, the late sixties to early seventies was a time of anguish and growth, a time to reassess society and also to reevaluate personal relationships. I came to understand that it was *my* hangups that prevented me from entering a successful relationship with a woman. I was too busy propping up my own insecurities and manipulating others in order to maximize my sense of well-being. I finally had to look at what I was *doing,* not what I was *saying*. I was so insecure that I clung to relationships that weren't going anywhere, until I found someone else. These overlapping relationships were dishonest on my part and often unnecessarily painful. If I really wanted a good relationship, I would have to stop trying to maintain a lady-in-waiting and learn to be by myself without panicking. Seven years after my marriage broke up, I finally reached that point. It was then that I made a quantum leap, when I met Tara.

December 10, 1971 marks the beginning of a major transformation in my life. On that day, I gave a talk at Carleton University in Ottawa. When I entered the lecture hall, I found an overflow audience of over four hundred people. As I began to speak, I noticed a striking woman seated close to the front. When the lecture was over, a number of people came up to continue the discussion, and hovering at the outer edge was this incredibly beautiful woman. I finally announced (ostensibly for everyone, but mainly for her), "I hope all of you are going to the party after the panel tonight."

That night I participated in a panel discussion on "The Social Responsibility of Scientists." And sitting in the audience was the beautiful woman again. But in the hubbub after the discussion ended, I lost sight of her and caught a lift to the party with the other speakers.

When I entered the house where the party was, I was immediately engaged in conversation by a group of people, but was happy to see that stunning woman drift into view. Everyone wanted to talk about profound issues like science and society, while all I wanted to do was go over and talk to her. Finally, I looked past the circle of people, blurted "Would you like to dance?" and bolted for the living room, hoping she would follow. Apparently she looked at everyone else and asked, "Who was he talking to?" "I think it was you," she was told, and so she followed me out. The rest, as they say, is history.

The spectacular woman was Tara Cullis and, coincidentally, she was from British Columbia. Her parents lived in Squamish, about thirty miles from Vancouver. She had graduated with an honours degree in English from the University of British Columbia and had come to Carleton to work on a Master's degree in Comparative Literature.

Tara had attended my lecture in part out of curiosity and in part to support a fellow British Columbian. It was a decision that changed both of our lives forever. I've never believed in love at first sight, but it was clear that there was an immediate and powerful attraction for both of us. I've always felt I talked myself into her life because she listened to my speech first.

I flew back to Vancouver the next morning, but all I could think of was Tara. I waited impatiently for her to finish her course work and teaching so she could return to Vancouver for the holidays. By New Year's Eve I had asked her to marry me, and although she managed to hold me off, we were married on the first anniversary of our meeting.

What was it that had affected me so quickly and profoundly? The first thing of course was her beauty. But it was Tara's mind that overwhelmed me. She is brilliant. At twenty-two, she had been thinking about life with a sophistication that far exceeded mine. Her undergraduate honours thesis had been on Malcolm Lowry and I spent hours listening to her talk about her ideas on literature. Tara's father and brother were physicists and so she had a lot of science in her background. Consequently, I was able to talk to her about my interests and she expanded on them with her own background in the arts.

But perhaps people least appreciate the quality that is so important to me: her sense of humour. In my upbringing, life was tough, life was serious. Humour and jokes were not a big part of my culture. Tara taught me to laugh, not just at the world, but at myself. I still find it unnerving when I get into a real snit and start ranting only to have Tara laugh and tell me to "grow up." But she's very soft-spoken, and it amazes me to see men simply ignore her or talk louder over her so that they miss her wit and intelligence.

When I began to court Tara, I predicted that her parents would object to me for three reasons: my race, my age and my divorce. In fact, they were only concerned about my divorce. I had never met two people who were more colour-blind than Harry and Freddy (for Freda), Tara's folks. They looked me over very carefully as any parents should, but only out of concern for Tara's future, not because of any prejudice about Asians. I was astounded at how freely I was accepted.

Family ties are extremely important to both Tara and her brother Pieter. Their closeness to each other had been threatening or intimidating for many of their dates in the past, but for me it was like finding a second family. After Tara and I had been married a few years, we found a dream home. It cost more than we could afford and I felt no hesitation in inviting her parents to buy it with us. Today they live upstairs from us and it has been one of the best things we've ever done.

On the other hand, Tara found herself thrust into a difficult new role. She was the prospective daughter-in-law of someone who regarded "the English" as the enemy. During our engagement, there were many times when Tara had to endure my father throwing out accusations of "you English" or "your people." But Tara hung in there the whole time. It was only after I finally had a big scene in which I pointed out to dad that just because he had suffered from prejudice he had no right to be a bigot himself, that he began to change.

It was even more challenging with my children. Tara appeared as the proverbial "stepmother" with all of the connotations that carried. Although Joane and I had been split up for seven years, Tara was the first person who represented an absolute end to the marriage. Tara and I were very much in love, so I assumed the children would share in my happiness. But, naturally, their loyalty was to their own mother.

The girls accepted my new wife well, although I suspect that Tara was a pretty formidable role model. But Troy was a different matter. He was overwhelmed by the weight of being my son, and feeling pressure from his teachers and fellow students to live up to his dad. He had all kinds of things going for him — he is good-looking, an outstanding athlete and bright, and has a good sense of humour — but I was a continual reminder of the load he had to carry. And I was not as active a parent as I wanted to be. I couldn't lay down the law because I didn't live with him. He often interpreted an innocuous enquiry I might make about school or future plans as inordinate pressure. There were times when I was so frustrated at not being able to discipline him, that I felt my son and I were losing our relationship. To Troy, Tara was just another barrier.

But it was always Tara who defended him fiercely. She invariably took Troy's side when I talked about him and argued that I should try harder. Just as my mother had influenced dad to apologize without my knowing her role, so Tara had a profound effect on me with my son without his knowing.

During our first year of marriage, Tara and I were inseparable. We travelled together, and when I was an exchange scientist that summer in the USSR we took four months to go around the world visiting exotic places like Siberia, Nepal, Thailand, India, Korea and Japan. To me, it seemed like a fairy tale. But after a year, Tara decided that she had to continue her own career. She had been raised by parents who instilled in her a strong sense of her own worth and she decided that she had her own aspirations that demanded fulfillment. She wanted to go on for a Ph.D. in Comparative Literature and try for an academic position. There was no Department of Comparative Literature at UBC and we both wanted her to get the best education possible, so she applied to a number of schools and settled on the University of Wisconsin in Madison.

Tara's decision to go to Madison was not easy. I understood and supported her need to fulfill her goal of being a scholar in literature. Had I been her age, I'm sure her intelligence and ability would have severely threatened my male ego. But being older, I could revel in her abilities and feel proud that she finds me worthy of being her mate. The fear for me was the danger to our relationship of a long separation. At one point I was moping around the office when my secretary, Shirley Macaulay, butted in. As I moaned about how hard it was, she fixed me with that all-knowing look and said, "Look. Right now, two or three years seems like a lifetime to you and Tara. But in a few years it will be nothing." She proved to be absolutely right.

Tara went to Madison and put in two years as a graduate student. Contrary to my fears, the separation solidified our relationship. One of the first things we agreed was that we would talk to each other *every* day. We stuck to that and stayed in daily touch. The phone bills were astronomical, but insignificant when compared with what they preserved. Those phone calls became the focus of my days. I longed for them, saving up stories and little tidbits to relate. The calls became the high point in our lives and to this day, through all of our travelling, phone calls remain a critical part of our relationship. In fact, sometimes when we are together and get lazy about communicating, we tell each other we'd better go to different rooms and call each other.

SABBATICAL LEAVE

University faculty are eligible for a sabbatical leave every seven years. To the lay pubic, a sabbatical leave seems like the height of privilege. I believe that a sabbatical is crucial for the maintenance of a vital outspoken community of scholars. Scholarly activity often takes a back seat because the daily demands of teaching and administrative duties must be met first. Sabbaticals free academics to pursue their research and rejuvenate themselves with new ideas. As a rule, professors take sabbaticals in different places, even other countries, where they gain new ideas, perspectives and colleagues. But when times are tough and universities undergo close public scrutiny, privileges like sabbaticals become vulnerable to attack. It happened to me.

As I gradually became involved in broadcasting, I acquired a high public profile, something not common among academics. Since 1962, I had done work for radio and television in a sporadic way. I never thought of a serious involvement with television until 1969 when my first proposal for a science series was accepted by the CBC. Fortunately, I also received the E.W.R. Steacie Memorial Fellowship, which is awarded to the top scientist in Canada under the age of forty. The fellowship paid a faculty member's salary so the university could hire a replacement and free the award winner to work at research full time. It's an excellent idea to encourage good research. For me, it coincided with my first heavy involvement with television.

Academics always have a difficult time assessing a "vulgar" medium like television; they operate in a world in which their output is carefully scrutinized by other scholars. So we try to be as precise as possible and qualify potential ambiguities or speculation. We don't like to say "it is so"; we prefer to write "it appears" or "the data seem to indicate" or "it is highly probable." In television, scripts have to be more direct. Programs must make generalizations that often ignore rare exceptions, and long interviews with experts may be edited down to a twenty-second snippet. Academics don't like any of that.

My growing involvement in television was resented by my fellow professors. I can't second-guess all the reasons, and there were undoubtedly many. I heard some of them: I was on an ego trip, my science wasn't good enough so I shifted areas, I was wasting my time. I felt the disapproval by my colleagues came mostly from their view that popularizing science through broadcasting was beneath the dignity of a university professor.

I was hurt and stunned by the reaction of my colleagues. What upset me most was that I only heard what they felt second or third hand. No one ever said to my face, "You shouldn't be on TV" or "You're just a grandstander." Those were the things my students heard. I had not involved my students in my television work and I knew many of them disapproved of it themselves. I felt it was totally unfair for them to be put in the position of defending me.

In 1974, when Tara had decided to go back to graduate school, we knew we would be separated for at least two academic years. One way of minimizing the separation would be for me to spend a sabbatical year at the University of Wisconsin. There was an excellent group of geneticists there, so I applied for a sabbatical leave to be spent at Wisconsin. When a sabbatical is granted, the university pays 60 percent of a year's salary while the professor finds the remaining 40 percent from other sources.

I had been solicited by someone from the Killam Foundation to apply for a grant. The Killam family left an enormous fortune to be distributed in prizes and fellowships, but initially the Foundation wasn't well known and didn't receive enough qualified applications to use up the annual interest. So I was practically guaranteed an award. I applied for a Killam in order to go to Madison and work on a genetics textbook for nonscience majors. I was shocked to be informed very late in the year that I had been turned down for the Killam fellowship. (Years later I found out that an eminent human geneticist from McGill University had objected vehemently to supporting me because the book I was writing might be critical of medical genetics.)

I was left without a Killam grant and therefore missing 40 percent of my salary while Tara was enrolled as a foreign student at the University of Wisconsin, which was itself a considerable financial burden for us. It never occurred to me to forego the long overdue sabbatical (I'd waited twelve years instead of seven) and stay at UBC. I wanted to be near Tara, so I decided to use my sabbatical to move to Toronto, where I could continue my work as host of "Science Magazine," and more easily commute to Madison. I told the head of my department about my intentions. Meanwhile, I had always negotiated contracts with the CBC on the basis that so long as my total income was about equal to my year's salary at UBC, I was happy. That was the basis of my agreement during my sabbatical year.

However, a battle ensued over my sabbatical leave that was precipitated by Doug Collins, an English war hero who had achieved notoriety for his outspoken views on television and in newspapers. Collins, by the seventies, was known to harangue the public through his column in the *Vancouver Sun.* His reputation was built on his views and a pugnacious style that seemed admirable when he was attacking government and big business, but increasingly he focussed on the visible minorities, ethnics who seemed to be diluting the British WASP society that Collins treasured. His columns took on a racist and anti-intellectual direction. He often attacked the university as a symbol of elitism and privilege and left-wing do-gooders.

In a column attacking universities, Collins concluded a half-page article with a paragraph asking why the taxpayer should be paying money to me while I was working with the CBC on my sabbatical. His previous attacks had already put the vulnerable university administration into a defensive mode and this comment must have sent them into a frenzy. Suddenly, they *discovered* that I was not at the University of Wisconsin as I had indicated in my initial application for sabbatical. But they never bothered to confirm that I had informed my head of my change in plans. They also assumed that I was being paid by the CBC at a level that far exceeded my UBC wage. As a consequence, I was informed that my sabbatical was being revoked. I was astonished and outraged at the *ad hoc* manner in which they made their decision — no facts, no personal interview, no consideration of right or wrong.

Shortly after I had been informed of the revocation of my sabbatical, I was giving a lecture at the Ontario Science Centre in Toronto. In the audience was Lydia Dotto, a science writer for the *Globe and Mail.* "Don't you encounter difficulties with your fellow faculty members for your attempts to popularize science?" she asked during the question period. I replied, "You bet! As a matter of fact, UBC does not consider my work with the CBC proper academic activity and has just revoked my sabbatical." I didn't know then who Lydia was or for whom she worked. I soon found out. The story of UBC and my sabbatical made the front page of the *Globe and Mail* the next morning.

The administration was now forced by their kneejerk response to Collins to press their case. Once I had made my point that I had informed the head of my department of my change in plans, and that I was not profiting from my leave, the administration retreated to the position that popularizing science through television was not proper academic activity for a sabbatical leave. It's worth noting that of the four administrators involved in the controversy (president and three vice-presidents), three were scientists.

I was so angry and disillusioned that my inclination was to quit the university. At that point, an extraordinary thing happened. The controversy drew a public response which was overwhelmingly favourable to my activity. I was one of the few university people who had a public profile and lay people called and wrote to show they approved of it. Scientific organizations, such as the Biological Council of Canada and the Canadian Society of Zoologists, passed resolutions supporting me. UBC was hammered for its shortsightedness.

Most faculty members at UBC treated me like a leper. They simply didn't want to get caught in the issue and ran for cover.

It raised a lot of embarrassing questions about what people really *do* on their sabbaticals, and how much they are paid. About this time, two remarkable men — Tom Perry and Jim Foulks of the Pharmacology Department — came to my aid. Both are ex-Americans who had left the United States during the McCarthy era and have long been known as defenders of matters of principle. They were also active in the Faculty Association and recognized that the adminstration's actions were wrong. Foulks and Perry negotiated with the adminstration on my behalf: the administrators had made a mess, they said, where none had been necessary, and now had to extricate themselves.

In the end, I was not only granted a continuation of the sabbatical, I was informed that the university had never meant to criticize my efforts to popularize science. On the contrary, the university was very supportive of what I was doing. "Is there any way we can help you?" one of the vice-presidents asked, all sweetness and light. "Yes," I answered immediately, "I would like to take a leave of absence from the university for two more years." It was granted. I had lost my innocence about what university and its faculty stood for. Academics and administrators are just as shortsighted and craven as any other group of people, but it was the Perrys, the Foulks and the public who reinforced my faith in people and gave me my first inkling of the real support there was for my television programs.

In the years that followed, television became my major commitment. As the bright promise of Flower Power faded and economic recession set in, students became more concerned with jobs and security than with social activism and university reform. I stopped teaching genetics for science students and devoted all of my formal teaching to a course for nonscience students.

In the late seventies, I began to receive strong, positive support from scientists for my programs in radio and television. As fiscal restraint became a government priority, science, perceived as a frill, became an area to cut. It was when Prime Minister Trudeau's budget cuts hit the already woefully underfunded area of science, that scientists realized the necessity of public awareness and support.

Eventually, I renegotiated my relationship with the University of British Columbia; I did not want to sever my connection with academics. As a member of the university, rather than an ex-professor, my criticisms of the academic community are considered more seriously by faculty. In turn, so long as a university supports me, I feel more strongly about defending it as a place that tolerates a broad range of people and ideas. But I could not work full time on campus and remain as involved in my television work.

Today, I am no longer a full-time faculty member and I no longer teach my course at the university, although I do give guest lectures. I am now supported by the Department of Community Relations and see my role as a representative of the university. I speak throughout the province to high school students and produce short audio and videotapes about UBC. But I've never felt I have stopped being a teacher. The television medium I now use is vastly different from a classroom performance. I have all of the props of electronic wizardry, but I no longer have a captive audience. My viewers deliberately choose to watch my programs; they're not satisfying any curriculum requirements. While the number who watch each week is gratifying, television is sometimes a frustrating medium — ephemeral and unidirectional. I can't see the people I'm trying to reach, but I know they're there. I meet them all across Canada.

A session with students at Carleton University the day I met Tara Cullis, December 10, 1971.

The children grow up. (l to r): Laura, me, Tamiko and Troy.

Tara Cullis.

December 10, 1972. Our wedding (l to r): Dad, Mom, chubby me, Tara, Freddy, Harry.

With a DNA molecule outside the Biology Department at Cal Tech in 1972.

Tara's first successful fly-fishing experience under Dad's tutelage.

Doctor Tara Cullis receives her Ph.D. from the University of Wisconsin in 1983 just before the birth of Sarika.

A ling cod caught at Tara's parents' cottage. That's Pasha, their dog.

The biggest fish I've ever caught, a twenty-six-pound spring salmon hooked by accident while jigging for cod.

CHAPTER EIGHT

BROADCASTING — A DIFFERENT AUDIENCE

I DID NOT DELIBERATELY SET OUT to become a media person, but the zigzag path of circumstances and fortuitous events in my life coalesced around that career. My first foray into television was in 1962 and little did I expect or want what TV eventually led to — it made me a celebrity.

I've always regarded myself as merely a messenger of important information, but viewers respond to the *people* on programs. On reflection, that makes sense. The impact of shows hosted by stars like Peter Gzowski, Barbara Frum or Roy Bonisteel is based on how their audience perceives them as people.

Television creates a sense of familiarity and intimacy: we bring it into our homes, often our bedrooms, and get to "know" the people on the screen. People who appear on a single program flit through public consciousness and disappear. But repetition and staying power generate recognition — a public profile. It still shocks me when total strangers greet me like a close acquaintance.

Of course, it is the support of fans that keeps our programs on air. However worthy a show, it will die if it doesn't attract an audience. And over the years, those viewers have had another unexpected effect on me; they have given me a platform to disseminate my ideas. Because of my public visibility on television and radio shows, I am able to write regular columns for newspapers and journals and send off articles to different magazines. Were I an unknown freelance science writer, those avenues would not be as readily accessible.

Another effect that I've only recently come to appreciate is that now that I have been broadcasting for two decades, my audience has become a very real constituency. When I make a statement in public, implicit in it is an audience that "backs me." This is exactly the opposite of what I had set out to do. My hope was that through science programming, I would educate the audience, thereby empowering *them* to act. Instead the audience, in fact, has empowered *me* to speak out and be heard by people in high positions.

I have gone from being a full-time research scientist and teacher to being a full-time broadcaster in the electronic and print media. This has created my greatest personal pangs. After all, I was profoundly moulded by my training as a scientist, and everything in me itches to continue publishing scientific papers.

Long after my broadcast career took over my life, my scientific self-esteem pressed me to "keep my hand in." I have published well over a hundred columns and articles in newspapers and magazines, yet last year I still felt a special thrill when I was asked to write an article for *The American Zoologist* because it is a *scientific* journal. And it still pleases me to be asked to speak by groups of scientists. Although they are no longer my constituency, I yearn to be thought of as a fellow scientist.

When I began my television career, I was still deeply engaged in research and understood very well the disapproval of my peers towards my broadcasting efforts. To scientists, doing science is by far the highest activity to be engaged in, and all else is less worthy. Scientists, like any other group of people, like to be portrayed positively and dislike any obvious or implicit criticism. I am also sometimes critical of my fellow scientists and that, to many, rankles.

But over the years, popularization of science has become accepted as a more respectable activity by scientists. I think the high quality of our programs, the wide public support of such work and the fluctuating government funding finally helped dampen critics from the scientific community.

I appeared on my first programs not so much out of conviction as out of curiosity. It was a lark. Eventually, when I realized what a powerful medium television was, I saw it as a means of educating people about the importance and implications of science. In the beginning, I was angry and frustrated at how poorly science was funded in Canada, and I wanted to gain more support for the profession. I believed that if we could explain the esoterica of science in a way that could be easily understood, the public would support science. And the public would in turn influence the priorities of politicians to become more generous to scientists and scientific research. Most of my scientific colleagues chose to go directly to the politicians. They felt they could score political points without having to resort to the much more daunting challenge of changing public attitudes. I have seen politicians come and go with frustrating frequency. Each had to be educated about the importance of science. I now believe even more firmly that a profound understanding and support of science by politicians will result *only* when science is as much a part of our popular culture and education as are history and literature.

Television also provided a vehicle to express my concerns about the world and technology. It pulled me away from the limited perspective of a research scientist. Once I left my lab, I could see the enormous social consequences of science, its tight linkage with profit motives of private industry, its terrible dependence on military support. I recognized that the fruits of science pervade every aspect of our lives, but that science is so cloaked in jargon it remains hidden from the public. It became my faith that through greater public understanding and awareness, the *application* of science would come to be determined as much by public priorities as by military and industrial needs. Nuclear war, environmental degradation, social manipulation and control — these became matters of great concern that I could discuss through the media and hope to make an impact on people. Once involved, I couldn't go back.

GENETICS AND SOCIAL RESPONSIBILITY

As I learned more about the history of my professional discipline, my broadcast priorities gradually changed. My sense of social responsibility as a scientist increased as I became shamed by what genetics, my great love, had led to earlier in the century. Let me explain how I arrived at that point.

From the first day I began to teach, I thought it important to understand my students' interests and needs. I learned to speak to them without a lot of technical jargon about issues which were relevant to them. At the University of Alberta, those discussions with my first class fanned an interest in cloning farm animals and genetic engineering of domestic crops. Many of my students at UBC were pre-meds who engaged me in debate about pre-natal diagnosis, cloning humans and gene therapy. The results of those conversations often ended up in my lectures and students loved it.

As my campus lectures increased in popularity, I began to get invitations to speak to groups off-campus. I found myself extending my horizons by looking up material of interest to those outside groups. For example, when I talked to a group of *Nisei*, I considered racism and the question of hereditary aspects of behaviour. Addressing a church group, I looked at cloning, test-tube babies, abortion and the definition of human life. When a Hadassah group asked me to speak, I boned up on the Holocaust and discovered that Josef Mengele, the infamous Angel of Death, called himself a geneticist.

As my reading took me further away from pure genetics, I discovered a whole history of the science that had not been a part of my education and training. I discovered that even before the modern phase of genetics had begun in 1900, one of Charles Darwin's cousins, Francis Galton, had been interested in the inheritance of genius. He studied family pedigrees and "showed" that ability in music, writing, art and science ran in families and concluded that these talents must be hereditary. Galton became a champion of "eugenics," the science of improvement of human beings through selective breeding.

By the turn of the century, when the science of genetics was established, there was already a climate of thinking that most human behaviour and ability are genetically based. The laws governing physical inheritance in fruitflies and plants were found to hold equally well in mammals, including humans. In view of the apparent universality of genetic principles, people began to extrapolate freely from other organisms to people. But in their intoxication with their discoveries, geneticists began to expand far beyond legitimate boundaries of their data. They began to confuse their beliefs and value judgements with scientifically proven facts.

Edward East, a Harvard professor who became president of the Genetics Society of America in 1937, declared in 1919, "The Negro is inferior to the White. This is not hypothesis or supposition; it is a crude statement of actual fact." But "inferior" and "superior" are not scientifically meaningful descriptions — they are value judgements. Human behaviour is an expression of a complex interaction of heredity, physical make-up, personal experiences and environment. Very few behavioural conditions are inherited as a simple one or two gene-controlled trait. Moreover, the early theories about the inheritance of mental traits were often based on the heredity of *physical* characteristics such as feather pattern in chickens or fur colour in guinea pigs.

However, eugenics was accepted by the scientific community as a legitimate discipline. It was taught as a subject in universities; Eugenics Societies of lay and professional people were formed; and textbooks and journals were published. Through the principles of genetics, there seemed to be a rational basis for explaining the vast variation in human appearance and behaviour. Complex human traits such as nomadism, criminality, drunkenness and temperament were asserted by some geneticists to be inherited. Going a step further, often racial or certain socio-economic groups were identified as exhibiting some of these unwanted characteristics. The pitfalls were legion.

The scientists making claims about the inheritance of human behaviour in the early 1900s were not third-rate intellects grandstanding for attention. They were some of the leaders in the field who believed they were motivated by the highest humanitarian ideals. They saw their science providing the means of improving on nature to produce superior human beings, but they failed to understand the degree to which their own prejudices affected their objectivity. The history of science reveals that scientists are first and foremost human beings whose biases, foibles and limitations are carried over into their science.

Improvement of the human species was envisioned through two approaches: positive eugenics was achieved by encouraging people judged to have superior qualities to have more children than "mediocre" people. Negative eugenics was the converse; that is, discouraging inferior people from reproducing. The goals were thought to be humane — to rid humankind of the misery of mental deficiency, physical impediments, poverty, crime, sloth and immorality. But the science was atrocious.

As I absorbed the implications of this astounding period in the short history of modern genetics, I began to see that the two great passions of my life — genetics and civil rights — had become grotesquely intertwined. It had been geneticists who, by their grand claims, had set the stage for terrible acts of discrimination that ended in the Nazi Holocaust! Eugenics had been invented and spread by geneticists and had swept North America.

In the United States, members of the upper classes sought to overcome the threat posed by "the masses" by encouraging their families to have many children. In the United States and Canada, these pressures eventually led to the enactment of federal and state legislation to restrict immigration of people deemed to be inferior (such as "Mediterraneans" and "Hebrews"). American states passed laws prohibiting miscegenation (interracial marriage) while *Eugenics Acts* permitted sterilization of inmates in mental institutions in both Canada and the United States. German intellectuals and leaders followed the North American activity with great interest and admiration. When the Nazis imposed their *Race Purification Acts,* they modelled them after the American legislation.

When U.S. General John DeWitt, the man in charge of the Japanese-American evacuation after Pearl Harbor, had argued that Japanese born and raised in America were still tightly bound to Japan by virtue of blood ties, he was simply reiterating in popular language the claims so extensively propounded by geneticists: "A Jap's a Jap, no matter where he's born. They're sneaky and can't be trusted"; "Indians are lazy drunks"; "Blacks have low intellects"; "Jews are interested only in money" these statements of bigotry had been given support by the claims of the scientific community under the banner of eugenics.

The realization of the terrible consequences of scientists' well-meaning claims was horrific for me — I could no longer view science simply as a noble activity of a civilized society. Science had granted apparent legitimacy to some of our basest bigotry. It was a painful admission to make about a profession I loved. If I and other geneticists ignored the history of our profession, what could prevent a repetition of past horrors? I looked to television to make people aware of that historical burden my science carried.

Up to a point, I had learned to rationalize my research activity and my social responsibility as a scientist. It took a student to set me straight. In 1966, Martin Greenall, one of the brightest students I have ever encountered, took my introductory course. Marty invited me to give a lecture at his dorm. I gave a talk about cloning and genetic engineering and warned of its potential for good and bad. There was a lot of interest and a lively discussion followed. At one point, someone asked, "If you're so concerned about the misuse of genetics, how can you continue to do research?" I had been asked this many times before and I slid glibly into the answer. "Look," I said, "I am discharging my responsibility by demystifying what is going on in my area. Once *you* understand, you have power to decide for yourselves where it should go." I went on in the standard answer of a scientist: "I do *basic* research with fruitflies, not humans. My work has no direct or immediate application. I'm merely trying to expand the horizons of knowledge."

At that point, Marty stood up. "I don't think that's a justification," he countered. "All scientists contribute to a common body of knowledge. You contribute through your basic research.

Out of that pool of information, someone else may be able to extract an idea to construct a weapon or do something bad. And no one can disclaim responsibility if he's contributed to that totality of knowledge." He had skewered me with this simple, perceptive insight, and it shook me badly.

My initial answer to the student's question had been so easy and comfortable. That rationalization had allowed me to continue doing what I had always loved — basic research — without thinking about it. But with Marty's objection, I had to reexamine my assumptions. I felt frightened by the awesome responsibility that doing research now seemed to present. I was paralyzed by a reluctance to contribute any more to a body of knowledge whose potential for misuse had become so clear in my mind. Because the major users of scientific knowledge are the military and private industry, power and profit are the main determinants in the applications of science, so that questions of environmental impact or public health become subordinated. I didn't want to pursue the implications of Greenall's insight, but he had opened a door that I could no longer keep shut.

For about a year, I simply stopped doing research. Work in the lab continued, but I was sleepwalking my way through it trying to decide whether I could go on. In the end, I went back into the lab with yet another rationalization — this time more personal and real. I reasoned that if I wanted to continue my obligation of discussing the hazards of scientific application, I first needed to remain sane. If I were to go off the deep end, no one would pay attention to me. My first marriage had broken up and my personal life was in a shambles at this time. The lab was my only sanctuary; it was the one place I could go and have my stresses fall away as I immersed myself in science. It kept me sane.

That was my justification, but it turned out that by continuing to publish and thereby maintaining scientific credibility, I was buffered from being dismissed by other scientists. It allowed me to have a greater effect later as a broadcaster, because I myself was still an active member of the scientific community.

THE MEDIA — NEWS VS. TRUTH

I received a research grant from the National Cancer Institute for my work on the study of cell division. My interest was in the genetic controls over the complex events leading up to the division of a cell, so it was easy to draft a grant application in which this work was made relevant to cancer. One of the enlightened programs of the NCI is to bring science reporters together where they can hear scientists themselves talk about what they are doing. As one of the NCI grantees, I was invited to Toronto to talk to reporters. I discussed my work on temperature-sensitive mutations in *Drosophila* and was listened to with polite interest.

During the question and answer period, Joan Holloban, one of the top medical reporters for the *Globe and Mail,* got up and asked me a question about research and social responsibility. I told her about the dilemma I'd faced through Marty Greenall's question. In the stories about the day-long conference that were printed the next day, it was *my* anecdote about social responsibility that dominated the reports. I was shocked and upset that the science hadn't been the main news. The surprise for me was that the report revolved around the anecdote, a story that was not really the centre of the activity of the day. And that's the way news usually is. It is the personal human drama that interests reporters and readers, not the intellectual rigour or cleverness of the research itself.

There is a very real distortion in reporting. Today, the pressures on scientists to publicize their discoveries forces them to publish their work in dribs and drabs. In the early part of this century, genetics papers were exhaustive compilations of results — they were tomes that often became classics. Today, we may publish a paper based on a single result or observation; the vast bulk of current papers are small incremental additions to the scientific literature. But the scientist with a saleable angle — the isolation of DNA that affects an interesting trait or the identification of the chromosomal location of a specific gene — which is not in itself a major discovery, can get reported by the press because there seem to be long-term practical implications. "Breakthrough'' has become a most devalued word by its excessive use, and the public is becoming understandably skeptical. Poor press reporting distorts the nature of scientific progress.

Distortion of news can result in other ways. A severe restriction reporters must deal with is the limitation of space or time. A newspaper reporter who has to summarize an all-day meeting or a two-hour lecture in a six-inch column cannot possibly cover it all. And when television news items vary from fifteen to ninety *seconds,* the superficiality is even greater.

I have often had the experience that after delivering an hour's lecture on a topic such as ''Science and Social Responsibility,'' followed by a half-hour discussion, the reports that finally appear on it have nothing to do with my main topic. Often during the question and answer period, I may be asked about something unrelated to the talk. So, for example, someone may say, ''You're a geneticist. Is cloning possible?'' And when I give a flippant answer like, ''Yes, the technology to do it is there. I just hope our prime minister doesn't try it on himself,'' chances are it will end up as the report. Often it will be headlined sensationally: ''Suzuki warns PM against cloning self!'' And the innocent reader in St. John's or Victoria concludes my speech was all about cloning.

I have a special aversion for headlines. Writing up headlines is a very demanding art and is done by people who specialize in it. They don't write the articles. In 1971, Sandy Ross wrote an article about me in *Maclean's* magazine. In one part of the article, I was discussing the ramifications of genetic engineering. The whole article documented my concerns about the enormous power and danger of the technology. He correctly quoted me in this context as saying, ''Make me a dictator with the power to say who mates with whom and I could give you a race of people you wouldn't recognize.'' I went on to warn of the hazards of this kind of thinking. The headliner wrote, ''Make me a dictator, and in three generations I could give you a race of people you wouldn't recognize.'' All of the context of the remark was gone, and for years after, various people who hadn't read the article but remembered the headline would call me a fascist.

Holloban's article on social responsibility greatly increased my profile with professional reporters as someone worth paying attention to, and that raises another point. People in the media tend to feed on themselves. They keep files of reports on people, and as the file thickens, the first articles shape the ones that follow. So a lazy reporter will simply scavenge reports made by people before him. Often some error or exaggeration will be perpetuated over and over until it becomes "truth." This must be the bane of people in the limelight.

A NEW CAREER

When I took up my position in Edmonton, word of my lectures at the university must have gotten around. In the winter of 1962, Guy Vaughn called me. Guy was in charge of the university's programs in the electronic media. The university had access to the airwaves through a community television channel that had a time slot early in the morning on Sunday. "Your University Speaks" couldn't have had an audience much larger than the presenter's family and a few desperate insomniacs. Its format was strictly low budget, essentially a single camera focussed on a classroom stage with the option of a screen on which slides could be projected from the rear.

Guy contacted me about appearing on "Your University Speaks" to talk about any subject I wanted. I think he offered an inducement of twenty-five dollars but, of course, I would have done it for nothing. I can't even remember what I talked about, but Guy liked it. Most important, I found that performing in front of the unforgiving eye of the camera was not at all intimidating.

Guy came back to ask me to do more. I ended up lecturing for eight programs and I count them as the beginning of my career in broadcasting. At the time, I was not at all interested in pushing my way into this activity; it was just another experience. I was not a television watcher myself, so it didn't seem like a big deal. Science was far more exciting. When I moved on to Vancouver to start a new lab, I didn't think about television at all. But people sporadically asked me for interviews on radio and television, and over time reporters got to know me as a "good interview."

In 1969, at the height of the counter-culture movement, a highly placed CBC staff member in Vancouver got the idea for a radically different show. He brought together a disparate group of people — hippies, business people, academics and radicals. I was one of four people who were asked to lead discussions about the changes that were taking place in people's heads. The CBC executive's idea was to film the leaders of the sessions and all of the participants in the audience. Then we would go to another location and film the participants watching films of themselves. Then we would get back together in the first setting and discuss our reactions to the film and that would be filmed! You can see that this could go on ad infinitum.

Not surprisingly, it turned out to be a disaster. No one but the executive had any idea why we were there. We had a lot of fun, but it sure as hell was not going to be great television. Quite a bit of money had been directed to this idea, and the Vancouver people were embarrassed by not having anything to deliver for the four time slots they had committed themselves to filling.

It just so happened that at this time, Al Kapuler, a brilliant scientist from Rockefeller University, was visiting me. I had first encountered Al when I went to a meeting of the Genetics Society of America in Boulder, Colorado. There I heard the lecture he gave on his undergraduate honours thesis at Yale. It was a brilliant piece of work and earned him the highest grade Yale had ever awarded. It was eventually published as a paper in the *Journal of Molecular Biology,* one of the leading journals in the world at that time. The son of a New York Freudian psychiatrist, Al had embraced everything that came with the radical movement of the sixties — sex, drugs, anti-Vietnam War, civil rights — everything. I was instantly attracted to him for his charisma and brilliance, and we became very good friends. My relationship with Al was based on a genuine sharing of excitement over new ideas. We would exchange recent information and bounce ideas off each other, wildly speculating, theorizing, projecting. It was always great fun. He had come to Vancouver to visit me.

As a result of the little experience I had had with television, I thought it might be interesting to convey to a lay audience the sense of excitement when scientists get together to talk about new ideas. I proposed that Al and I meet in a studio where, in full view of the cameras, we would talk for the first time in months about some of the things that we were really enthusiastic about — jargon and all. My proposal came just when CBC Vancouver needed something to fulfill its commitments, and so it was accepted.

Al and I deliberately held off talking any science until the day of taping. You should understand that we both had hair down to our shoulders and were dressed in all the brilliant garb — bell-bottoms, dazzling shirts — the fashion of the time. We must have been quite a sight. We were big hams anyway, so when we got into the studio, we let ourselves go, drawing on the blackboard and talking a streak, oblivious to the crew around us. We went on long enough for three half-hour shows. When the material was edited down, I was called in and did a bit of narration to provide some explanation of what we were drawing and saying.

With those three shows I realized what a powerful medium television could be. I loved the *idea* of making science available to the lay person. Keith Christie, a local CBC producer known as a workhorse, had been called in to do the salvage job by filming Al and me. So I got to know Keith and suggested to him that there ought to be more science programs on CBC. Keith grabbed the idea. I now realize that it was not necessarily because he agreed with my rationale, but he must have seen some potential in me and wanted to use that to produce more shows. His interest was in production, so if I could be a vehicle, Keith was right behind the idea. Keith floated the idea up through the CBC brass.

As an innocent, I assumed that ideas were assessed on their merit and importance and that decisions were made after a lot of deep thought. I now know this is far from true. Personality, individual biases or preferences, power struggles and a lot of other factors weigh heavily. I had visions of everyone in Toronto going,

"Hey, how come we never thought about this before? Science! Of course. This is extremely important and we better give it a good budget and a slot in prime time." At that time, I didn't know that "The Nature of Things" already had a large and loyal audience. Although it had aired for over a decade, I had never seen the show.

Knowlton Nash, who had moved from being a top-notch on-camera reporter to a high-level post in CBC management, had the power to approve new shows. Prime time (which is from eight to eleven in the evening) was out of the question for us. Competition for those three hours is fierce and we didn't have any slick brochures, research reports or even a formal proposal — it was just an idea. Knowlton informed Keith that the only possible free time was at two o'clock on Sunday afternoons when most of the continent was watching football. Keith said, "I'll take it," his reasoning being that it represented a chance to broadcast to a national audience and we'd get our foot in the door. But it was a joke. When half-hour shows like "The Nature of Things" were getting perhaps $25,000 for direct costs, we got $1,600!

As we were gearing up to do the series, Keith got in touch with Knowlton. Nash asked, "How's the new show coming? You know, the Suzuki-on-science thing." "That's it!" said Keith. Knowlton was perplexed until Keith informed him that he, Knowlton, had inadvertently named the show — "Suzuki on Science." Once we had received the go-ahead, I discovered I was in way over my head. I knew absolutely nothing about television production. I wasn't familiar with television shows and didn't know how they were packaged. It's one thing to think, "Gee, we better do a show on genetic engineering," but a totally different thing to actually do it. My concept had been to find good scientists who could sit and talk intelligibly about their work for the lay public. And essentially that's what "Suzuki on Science" was — "talking heads," which is anathema to most television producers.

All of our shows were shot on tape because that way the

shooting didn't come out of our direct costs budget. We shot with huge equipment that took a lot of muscle power to push around. So whenever there was filming to be done, enormous trucks would arrive with crews, miles of electrical cable and a lot of gear. It was highly disruptive and since much of the filming was done in my office, television became an intrusion on the people in my lab.

Editing of tape was incredibly time-consuming. This was before the kind of computer-controlled electronics that make editing tape today a breeze. In order to "cut" a shot out electronically, two big machines had to be aligned. One carried the two-inch master tape and the other the copy. Since the machines didn't start up instantly, there had to be a certain allowance of lead time on the tape before the shot so the rolls could get up to speed. Technicians would go back and forth trying to get both tapes synchronized, so that just the right clip would be copied and transferred without any blank space between the shot before. I seldom stayed in the editing room because the boredom would send my blood pressure sky high. If I have learned anything in all of the years of film making, it is patience. I've learned to daydream and think of other things while waiting for lighting, camera and sound equipment to be set up.

"Suzuki on Science" reflected my own interests and the available talent in Vancouver. We husbanded our meagre money rations to make one trip each year, one down the east coast and another down the west. We shot all of those interviews on film and nabbed a number of Nobel laureates and other eminent people like Jonas Salk. I don't know how we did in the ratings or whether the CBC even bothered with ratings in that time slot. But the show was cheap and we ran for two seasons.

Through the people I interviewed, my horizons were extended beyond the restricted blinders of my discipline. I could vicariously explore fields I wouldn't otherwise know. And if I do have a talent, I hope that it is as an interviewer. To me, Peter Gzowski and Patrick Watson are in a class all by themselves as interviewers.

They are intelligent, they listen, they are not afraid to follow up on something unexpected and they don't feel they have to impress the listener. I aspire to being in their league.

Once we were into our second season of "Suzuki on Science," I felt more and more frustrated at the limitations of what we could do because we had no budget to upgrade the quality of our shows. I didn't realize it, but Jim Murray, the executive producer of "The Nature of Things" was keeping an eye on our show as potential competition. It is a compliment that he took us seriously, because we simply could not match the quality of presentation of his show. Without any hope of increasing our budget substantially, I didn't see the point in continuing to turn out what were essentially radio programs. So although "Suzuki on Science" was renewed for a third season, I quit at the end of two.

To be honest, I hoped that when I quit "Suzuki on Science," someone in the CBC would come scurrying out and say "Gee, David, we agree with your goals and we don't want to lose you. What if we give you a better time slot with a bigger budget?" But this didn't happen. We weren't regarded as anything more than filler for a lousy time slot. We didn't have the money or talent to do anything exciting, but it didn't matter what we did as long as it wasn't disastrous. In fact, we did cover a lot of exciting and important subjects — women in science, aging, cloning, death, cancer and radiation were a few of them. But with our budget and time slot, we couldn't deliver a big audience — it was a no-win situation.

It had been an exciting two years for me doing "Suzuki on Science." I interviewed a number of scientists I would run into in later years and many became friends. I was also caught up in the most exciting phase of my research career, so television still definitely ran a poor second to science for me. I went back to the lab with no regret or any great disappointment with such a brief experience in national television. And for several years thereafter, I remained a full-time scientist.

PRIME TIME TELEVISION

In 1974, I met Jim Murray, the executive producer of "The Nature of Things," and my directions and priorities changed for good. But first let me give you some background on Jim. Jim knew he wanted to go into the electronic media since his high school days as a sports broadcaster. He trained in communication in college and first began working in radio before he shifted to television. He started working on "The Nature of Things" in the second year of its existence in 1960, and has been the guiding force behind the series for most of its existence. He is its heart and brains.

In the early seventies, the CBC had attempted to duplicate the enormous success of dramas like "The Forsythe Saga" by producing a blockbuster called "The White Oaks of Jalna." It was an expensive, lavish series that turned out to be a monumental dud. In the fallout from that debacle, it was decided to try to recoup some pride by producing a winner — Pierre Berton's best-seller, "The National Dream."

The person chosen to produce it was crucial — he had to deliver a ratings success. The search ended with Jim Murray. Most producers would leap at a chance like this because it could be a stepping stone to bigger and better things. But not Jim. "The Nature of Things" was his passion — he believed in what the series stood for and it reflected his basic ideals and philosophy. Jim had to be persuaded to become executive producer of "The National Dream"; he reluctantly agreed on condition that he would return to a fortified science unit when it was over.

The rest is legendary. Berton and Murray went on to turn out a monumental series that attracted huge audiences. It magnificently portrayed a sense of the breathtaking scope of the history of Canada. Jim did return to the science unit, but now Nancy Archibald, a creative and original producer, had taken over his old position of executive producer. As a reward for his great success on "The National Dream", Jim was given the go-ahead for a new series to begin on air in the fall of 1974. He would be executive producer of a thirteen-part series called "Science Magazine."

"Science Magazine" was to be a weekly half-hour collection of reports on science, technology and medicine. Where "The Nature of Things" looked in depth at one topic per show, "Science Magazine" was a series of short reports. Jim had hired a researcher, Richard Longley, to travel across Canada and find people and topics that could be used in "Science Magazine." He visited me to talk about my work, and for the first time I heard about the new science series. I was filled with envy that the Toronto people had been given the chance to do such a show after we had begged for more support to do a similar series. As soon as he left, I wrote to the CBC area head of science and asked whether there might be a role for me in the series. He never replied.

Richard had suggested that they might do a short report with me about different mutations in *Drosophila*. He proposed that the next time I visited Toronto, we might film something there. As a new season suddenly loomed up, Jim had to turn his attention to all of the picky details of packaging a new show. How could completely unrelated topics be sewn together so the transitions wouldn't be too abrupt? Jim finally decided that he needed a host.

In the fall of 1974, I took some flies with me to Toronto and Richard arranged for me to do a short on-camera piece there. It was set up in a lab at the University of Toronto. Jim was there to oversee it and fired questions at me. I was to answer, looking directly at the camera. I didn't know that Jim had decided to have a host. Indeed, I assumed from the lack of response to my letter that there was no future for me with the CBC.

But, in fact, Jim had been getting a lot of advice from geneticist Lou Siminovitch at the University of Toronto. Lou had given Jim a list of scientists with an evaluation of both their research reputations and their potential for on-camera work. Lou gave me high ratings for both. But I didn't know any of this. Although it wasn't intended as such, my interview turned out to be a fortuitous screen test. Jim looked at the footage and decided that I would do as the host of "Science Magazine." When Jim offered me the chance, I leapt at it. It was the best opportunity to be involved with high-quality television I had ever had. Fortunately, on short notice, my superiors at UBC allowed me to take a leave without pay and I moved to Toronto after Christmas.

I very quickly came to appreciate Jim's enormous talent. He continues to hold the science unit together with his low-key but demanding control. He is unrelenting in his attention to every detail in the production process. He is tough and uncompromising in his standards and he has an uncanny ability to look at a film and offer a detailed and constructive critique.

Jim has gathered around him a group of people who are committed to science and natural history films. By his integrity and loyalty to them, Jim commands wide respect. I came to love him as the best friend I've ever had. He taught me by example — by doing — and I learned far more about television in that first year than I had in all the other years combined. Right from the beginning, Jim trusted me enough to let me decide what questions I would ask scientists we were interviewing. He let me write all of the material I would say on camera. His faith in me gave me the impetus to do my best work.

He also gave me some of the most important lessons I've had. Rudi Kovanic is an award-winning cameraman who is the heart of our film crew; he's also one of the most meticulous, maddeningly careful cameramen you'll ever encounter. One time, I was poised to deliver my lines, as Rudi set up the camera, arranged the lights and organized the background. He fussed with this and that, checking his light meter and the lens focus over and over. Finally I couldn't stand it any longer and blurted out, "For chrissake, Rudi, let's do it!"

Jim immediately pulled me off the set and took me to a far corner. "Listen, Suzuki," he hissed, "everyone here is doing the best they can to make *you* look good. And believe me, with *you,* that's not easy!" I crept back on to the set like the chastised child I deserved to be. In truth, all of that effort culminates in what the audience sees on TV, and from that they form their opinion. When it goes well, *I* get the credit for it. So there were the cameraman, soundman, the lighting man and everyone else breaking their backs to make me look good and I was getting impatient. I've never complained since.

By the criterion of audience size, "Science Magazine" was a great success. It very quickly gained a large and loyal audience with a high proportion of young people, a distribution that was unusual for CBC. But before the first of the thirteen shows had even been broadcast, we learned that the series would be a one-shot affair and had already been cancelled for the following season. At the end of the last show of the season, I came on camera, announced the end of the series and said goodbye. The response was astonishing. Several hundred letters came in from people praising the show and demanding to know why it was being cancelled. "Science Magazine" was quickly reinstated and went on to run for five very successful years.

I loved working in Toronto because of the enthusiasm and quality of people who produced the programs. But there was something disconcerting about the medium of television itself. I felt the shows I was doing really were of a quality that could compare internationally. Yet all of the thrust of the production team pushed toward meeting deadlines for broadcast. Even before a show ran on air, it seemed that we had to start worrying about the next week's show. As each "item" was edited down, script written for it, sound effects added and narration recorded, its producer knew the report by heart. By the end, he or she could announce, "It works."

I was more concerned about whether it also works for a viewer who would not know it as intimately. What kind of impact does our show have on an audience that watches television in big blocks of time? How do we assess the effect the shows have? These were questions that have to be asked by all people involved in producing television. It especially bothered me because there seemed to be so little time or interest in posing them, let alone trying to get answers. Television fragments information into smaller and smaller units. Scenes shift quickly with some shots lasting less than a second. What we seemed to be doing with "Science Magazine" was whetting the appetite or titillating the viewer.

GETTING INTO RADIO

During that first season of "Science Magazine," I was asked by the UBC Alumni to give a speech in Toronto. I was happy to do so, and gratified when a large audience showed up. I spoke about my concerns about the importance of science in society and the need for an informed public. I didn't know it, but in the audience was a UBC alumna, Diana Filer, who was the executive producer of the CBC radio series, "Concern." Before that, Diana had produced "Gerussi," the morning talk show hosted by Bruno Gerussi that was the forerunner of "This Country in the Morning" and "Morningside."

Diana had put forward a proposal to produce a weekly one-hour radio program on science. When it was approved, she called me to introduce herself and asked whether I would be interested in hosting the program. Apparently my talk to the UBC Alumni had been like a test for the job and I had passed. My only experience with radio had been in short interviews done either by a reporter with a tape recorder or in a studio. I could see that radio was obviously less complicated than television, but because radio relies on talk and sound effects alone, it requires a large mass of material. Diana suggested that we do a pilot program to see whether we could work out a format and determine how it sounded.

I was intrigued and agreed to try. She had learned of a special media orgy arranged each year by the American Association for the Advancement of Science (AAAS). At the meeting, some of the most eminent scientists in the world are invited to discuss topics ranging from research in outer space to the latest cure for cancer. Scientists present papers and offer themselves for interviews. There may be as many as several thousand people attending and hundreds of scientists making different presentations. It was Diana's idea to go. I would carry around a tape recorder and try to interview anyone who sounded interesting.

The meeting that year was held in New York. I felt awkward and shy trying to buttonhole people, but soon found that the meeting was held just for that purpose and most people were flattered to be asked and extremely cooperative. It was a good learning experience for me, and set the precedent followed in all of the later years of the show. The annual AAAS meeting became the opportunity for us to stockpile dozens of interviews that could then be used throughout the rest of the year. We also did interviews around Toronto and in studio, and quickly accumulated enough material to "package" into a one-hour show.

I always have a hard time turning something down if it is interesting to me, so Jim Murray has to compete for my time with other projects that I tend to accumulate. He learned what was in store for him when I accepted the challenge of the radio show. After Diana and I had packaged the pilot radio show, I returned to Vancouver at the end of the season of "Science Magazine." I was all excited because my first radio show was to be broadcast and invited Jim over to listen to it with me. About halfway through the show, I became aware that my voice was starting to rise in pitch. Apparently something went wrong in the transmission and gradually the tape started to speed up. The last interview on the show was with a woman from the Toronto Metro Zoo and both she and I sounded like The Chipmunks. I was humiliated as Jim rolled on the floor laughing uncontrollably at the debacle. That's when I realized that he wasn't as excited about my radio show as I was.

In spite of the transmission problem, everyone was happy with the pilot and it was decided to go ahead with the new science show. Diana came out to see me in Vancouver. One of the first things she did was to go to the PR department at UBC to talk about potential interviewees for the show. The person she talked to scoffed skeptically, "You'll never get enough material to fill a weekly hour show." I soon found that we had enough material to do a *daily* science show if we had the resources.

It was Diana who named the show "Quirks and Quarks," and chose its theme music. I was a novice in radio and felt such a deep admiration for her, I was shocked to be told shortly before returning to Toronto that she had been promoted and wouldn't be involved with the show any longer. The new producer was to be Ivan Fecan, a twenty-two-year-old I had never met or even heard of. Diana had gone into hospital for an operation when I got to Toronto, and I immediately went to see her. She was just coming out of the anaesthetic as I rushed in. I asked her how she was and before she could answer, I launched into a tantrum about her abandonment. She says she was so groggy, she didn't know what the hell I was talking about. The fact is, I was scared stiff, and without her I didn't think I could deliver.

I shouldn't have been worried. Ivan was young, but he was ambitious. His career has been meteoric. He went from being the youngest head of Variety at CBC to a vice-president at NBC before he was thirty-five. He used to say to me, "Anne Murray was famous by the time she was twenty-five and I'm already twenty-two!" Ivan gave "Quirks and Quarks" everything he had. The show represented a big opportunity for him and he used it to create a brand-new way of presenting science.

In the first two years of "Quirks and Quarks," we tried all kinds of ideas, some of which worked, many which didn't. We had a weekly interview with an inventor; we had "Visions of the Future" which were dramatizations of predictions taken from old magazines. We answered questions sent in by listeners (many times we made them up). We had an actress who dished out homespun philosophy about modern society. We played actual recordings of people like Einstein and Freud. They were developed by Anita Gordon whom I liked because of her hustle. I later invited her to produce "Quirks and Quarks" which she did and she has stayed on with the show ever since. When I was hosting "The Nature of Things," I learned of Jearl Walker, a physicist who had written a book called *The Flying Circus of Physics.* I realized that the material in the book was perfect for "Quirks and Quarks" and suggested that we use Jearl on the show. We did and he too has been a regular.

"Quirks and Quarks" was an instant hit. In the beginning it was broadcast on a week night, and it drew a significant audience of some 50,000 listeners — minuscule in television and pop radio terms, but large for CBC radio in that time slot. The decision to move it to Saturday at noon, which is prime time for radio, was a reward for its success. The program was a smash hit and I have been proud to watch it continue to draw a large and loyal audience and garner numerous awards. Jay Ingram is an excellent host and I take pride in the fact that the show is still in a form that can be traced back to those first years when we started it.

Scientists loved "Quirks and Quarks." For one thing, they were far more likely to listen to radio than watch television. For another, the format of interviews is more serious. On radio, a scientist gets to dominate a report and has a good chunk of time, whereas television gives short snippets. Scientists regard radio as a "more serious" medium.

I loved radio. "Quirks and Quarks" was a program that I felt most "natural" doing. The technology was simple, quick and nonintimidating to an interviewee. We could set up for an interview and then let the tape roll with nothing more intrusive than a small microphone. It is relaxed and spontaneous. It was possible to inject genuine humour. We could roll miles of tape and cut out what we wanted. There was also a huge difference from television in material that could be covered. Television's great strength is its visual component, but this also limits the range of subject matter. Only a fraction of science lends itself to visuals since most of it is abstract ideas. So where television can only present a small fraction of the breadth of ideas in science, radio can cover it all.

Ironically, while the medium of radio is simpler and faster to produce than television, it became much more demanding for me. For one thing, we burned through material at an astonishing rate and all of the interviewing was done by me. For another, we quickly began to broadcast fifty-two weeks a year. Granted the summer months were repeats, but they still had to be packaged. With television, the number of original shows was limited, I wasn't involved in all of the items, and I could do a lot of the on-camera work far ahead of time. I could plan for weeks or even months when I could do other things. Radio was relentless and much less flexible. I found myself running back and forth between my television office on Bay Street and the radio building several blocks away. Moreover, each spring I would move back to Vancouver where "Quirks and Quarks" would have to be produced by a different team. It got to be complicated and difficult.

Hosting both "Science Magazine" and "Quirks and Quarks" was a full-time job. I couldn't possibly continue to delude myself into believing I was still an active scientist, so I withdrew my name from all requests for research grants. I had to spend more time in Toronto and knew I couldn't keep taking leaves of absence from UBC.

In 1978, Don Chant, an internationally acclaimed ecologist and one of the founders of Pollution Probe, was the provost of the University of Toronto. He offered me a perfect job — I would have tenure, give a course at night for one semester and would be on a one-third salary. I took another leave from UBC and as a trial, accepted an appointment in the Zoology Department at Toronto for one year. If it worked out to everyone's satisfaction, I would quit UBC and become a permanent faculty member at Toronto. U of T appreciated me for my work in the media, as UBC never had. But with the imminent possibility of my leaving UBC for another Canadian university, UBC finally offered me a comparable position as a "full-time professor at one-third pay." In the end, my love of Vancouver as a place to live led me to turn down the University of Toronto offer.

"The Nature of Things" and "Science Magazine" were put together by the same team of people. As the host of "Science Magazine," I always felt my show was a poor cousin to "The Nature of Things." Whereas "Science Magazine" might do as many as five items in a show, "The Nature of Things" could dig into a subject for an entire half-hour. Producers like to get their teeth into a subject; ask them to do a five-minute item and they'll invariably show up with a ten-minute report swearing there is no way to cut it down. So they much preferred doing "The Nature of Things." I felt they produced items for "Science Magazine" with less enthusiasm.

Nancy Archibald, as executive producer of "The Nature of Things," was finding it tough doing a lot of the administrative work and attending endless meetings. She was anxious to get back to her great love which is producing natural history films. Jim had become convinced that I was a net asset to the CBC Toronto Science Unit, not only as an on-camera person but because of my background in science and my membership in the scientific community. I kept vacillating about whether I ought to get out of television and try to get back in the lab. I worried a lot about whether I was very effective at what I was doing. Jim was anxious to keep me involved. So it was proposed that "Science Magazine" and "The Nature of Things" be amalgamated with me as the host of the whole thing. The show would be called "The Nature of Things with David Suzuki" and fill a one-hour slot. Jim would take over as executive producer while Nancy would remain a senior producer.

It was an exciting idea because it offered the possibility of a much more flexible program. We could devote an entire show to a single topic or we could do one in-depth report with one or two shorter stories. We could also do a number of medium-length stories. It was a major opportunity, but one that would be much more demanding of my time.

I reluctantly concluded that I would have to give up radio to devote much more time to the medium that has the greatest impact on the public. Besides, "Quirks and Quarks" had an established track record after four years and I felt confident that it did not need me to keep it going. I still take great satisfaction in having lifted it off the ground and demonstrating that science could be made exciting to a listening audience. So in 1978, I left "Quirks and Quarks" to become host of the newly formatted, long-running series, "The Nature of Things." My life became much less complicated, while the new format did give us greater programming flexibility. I became the beneficiary of the long history of excellent programming on "The Nature of Things" and gained a loyal audience which the show had built over the years. It was the best decision I've made as a broadcaster.

(above) Hollywood, January 1972: The set for two films on heredity and cell biology.

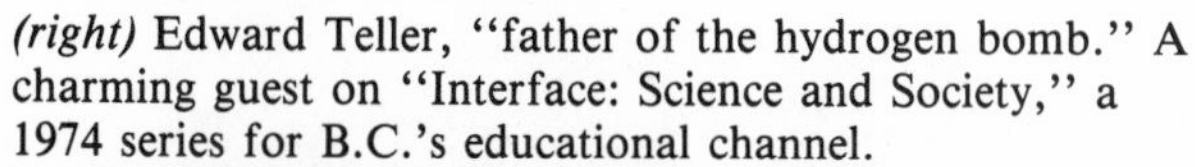

(right) Edward Teller, "father of the hydrogen bomb." A charming guest on "Interface: Science and Society," a 1974 series for B.C.'s educational channel.

(left) Starting the second season of "Quirks and Quarks" with the young, up-and-coming producer Ivan Fecan (1976).

(below) My best friend and Executive Producer of "The Nature of Things," Jim Murray. We had just landed on Coburg Island in the Northwest Territories to film our show on "Arctic Oil."

5

Nim, the chimp and his friend (1976). Nim was housed in a Columbia University mansion where he was taught to use sign language by Herb Terrace. 7

The award-winning film maker, Donald Brittain, playing a gas station attendant in the National Film Board film "This Is an Emergency" (1979). 8

An outfit for the National Film Board's "This Is an Emergency" (1979). 9

CHAPTER NINE

BROADCASTING UPDATE

A LOYAL AUDIENCE IN CANADA of 1.3 million views "The Nature of Things with David Suzuki" each week during the season. That's between 18 and 22 percent of the total television audience, an excellent showing against prime-time competition. Outside of Canada, "The Nature of Things" is the CBC's most widely sold program. Individual episodes are shown regularly in over fifty-five countries, while the entire series appears in thirteen, including the United States on the PBS network. It is one of the few CBC programs that, through sales, actually makes money for the CBC.

That track record is a tribute to Jim Murray and his production team which has kept up high standards for twenty-seven years. I also think that Canadians should congratulate themselves, because in the end they are the ones who have kept the show on air. Thanks to them, "The Nature of Things" has carved out a permanent niche on prime-time television.

When I started to work with the Toronto Science Unit in 1974, it was my long-term hope that CBC would make a major commitment to science programming in television by setting aside a specific time slot for science fifty-two weeks a year. The audience could then develop the habit of watching science shows every week on a certain day. Also, a year of shows would give us a hefty budget with which to try some novel ideas and to experiment a bit. But in times of economic restraint and budget cuts, this now seems an impossible dream. Science is still not a top priority area.

I think one of the burdens science programs carry is the perception within the CBC that our shows are "ghetto programs" — that is, they serve only a certain interest group. Too many people don't understand the vital role of science and technology in our lives today, or the fact that the public interest is there.

Once I went to see a CBC vice-president to plead for more time for science programs. When I told him science is far more important in our lives than business, politics or entertainment, he just stared at me incredulously. So I explained that at the heart of the issues of nuclear war, environmental pollution, energy, medical care and computer literacy, were science and technology. He replied "Those aren't science stories, they're current affairs!" His reaction reflected the fact that most of the people who rise through the CBC ranks come from a journalism background. To them, news and current affairs are all that matter. Not surprisingly, in all of the position papers written by the CBC, science programming has never been mentioned as a priority area.

To a large extent, the strong ecological perspective of "The Nature of Things" has been the legacy of John Livingston, now a professor in Environmental Studies at York University. John was once the executive producer of "The Nature of Things" and has continued to serve as a writer, narrator and philosophical guru for the unit. His ideas on humans and the environment are radical, running counter to the thrust of Western society. He sees humans as a species out of balance with the rest of nature, puffed up with an unrealistic sense of importance and incorrectly convinced of our right to exploit nature in any way we choose. I came into the unit with a human-centred perspective, revelling in our intellect and culture as special and unique. Livingston's ideas were a shock to me. I encountered them mainly through Jim Murray and reluctantly, over time, I came to understand the profundity of the "deep ecology" and green movements.

Livingston proposed a television series to Jim that illustrated, from his point of view, the place of humans in nature. About the same time, I had become interested in mythology as an expression of society's values, insights and wisdom. I was pushing Jim to consider doing a series to examine the modern mythology of our origins — creation as seen through science. I wanted to explore the way scientists imagine the origin of the universe and matter; how life arose on the planet; and the central place of DNA in life's diversity, development and death. I hoped we might compare the scientific world view with other cultures and ways of knowing. Jim responded by melding the two proposals into an eight-part series called "A Planet for the Taking."

The series stirred up a great deal of controversy and attracted an unheard-of-average of more than 1.8 million viewers per episode. Its critical success defied all traditional expectations. The audience found some scenes, such as the bull-fight and animal experimentation sequences, hard to watch. They often disagreed with our thesis, and many found some of the ideas difficult to follow. By all conventional notions, these factors should have turned people away. The fact that the audience hung in there week after week strongly attests to viewers' appetites for serious ideas. Too often people in the television business justify the production of pap with the rationalization that the audience won't watch anything demanding. Our positive experience with "A Planet for the Taking" suggests that we'd better reassess our assumptions about what the "public" actually wants.

As the on-camera host of a series, I am often credited by the viewing audience for the ideas and the production of the programs. That is one of the powerful illusions created by the medium. After all, the only indication of the complexity of production is the credits at the end of a program. The lack of public recognition is a source of irritation for those who have actually contributed the ideas and shaped the shows.

Although I acted as the host and narrator, it was Livingston's perspective that came to pervade the series and made it special. The three episodes for which he researched and wrote the script were provocative, challenging and original. Livingston deserves much of the credit for the philosophical thrust of "A Planet for the Taking." Jim Murray and producers Nancy Archibald, John Bassett and Heather Cook developed the programs, and with writer Bill Whitehead made it actually come together as a series. And I contributed as well.

Of course, I've left out the names of dozens of other people involved in the whole process of making a series, or any show for that matter. They do everything from researching background information and finding people to interview, to looking up "stock footage" which can be purchased from other sources, to all of the nuts-and-bolts things of checking the colour quality of prints, checking credits, timing the show, and so on. All of this complex business is hidden behind the on-camera person.

My own involvement with different reports and shows of "The Nature of Things" ranges from writing and delivering a short introductory piece, to spending a considerable amount of time interviewing people and making on-camera commentaries on location. One of my most satisfying contributions is to suggest story ideas to Jim, which are sometimes carried through and become shows.

The ideas conveyed in "A Planet for the Taking" deeply affected all of us in the science unit. After a series like that, we couldn't go back to the programming we had once done. Increasingly, our programs have questioned some of our most cherished notions of progress, the necessity for growth and the overriding concern for jobs and profit. We have since looked at acid rain, toxic chemicals in the Great Lakes, North American waterfowl, the Niagara escarpment and many other topics in a way that severely questions our ability to "manage" our natural resources. As a scientist, I know how ignorant we are of the biological and physical world, yet we continue to cling to the lie that we *know* what we're doing. The truth is we have no idea.

These days, with our vast choice of channels, we can find a show on wildlife almost any time of the day. Their popularity reflects the fact that the public, especially children, love nature films. Does this interest indicate some primordial need to connect with nature? After all, for 99 percent of our species' existence, we were not only embedded in nature, we *had* to understand it in order to survive. The camera provides new ways of seeing nature — shots from totally different perspectives, slow motion, at high speed or magnified. When you see a magnificent series like David Attenborough's "Life on Earth," you can understand how powerful television wildlife programs can be.

Even people who are in the business of producing wildlife films do not agree with taking an overt environmental stand. We all say that we are concerned about the fate of wildlife and make films in the hope of changing things. There are two views on this. PBS produces a lovely series called "Nature" that is hosted by George Page. In a symposium at the 9th International Wildlife Festival, George took the position that, by doing superb nature films, we chip our way into the public's sensitivities and, in order to maintain credibility, we must avoid taking an advocacy position. "The Nature of Things" represents the other pole. In our view, the media pour out stories that are full of assumptions and values in the guise of objective value-free reporting. Most programming on television simply takes for granted our right to exploit nature as we see fit, to dominate the planet, to increase our consumption, to create more economic growth, to dump our wastes into the environment. Few object to these assumptions because they are so deeply set in our culture that they are accepted as obvious truths. However, they are biases nevertheless. Yet the minute a natural history film takes a strong environmental position that questions these beliefs, it is immediately criticized and bombarded with the demand to present "the other side."

We can no longer afford the luxury of just doing pretty films. We have to take a strong advocacy position and point out wherever we can the error in our assumptions and values. The big problem we face with this approach is that this can become drearily predictable. The British refer to it as the "ooh, ah, oh dear" syndrome. What they mean is that we first see wonderful shots showing beautiful nature that make us go "ooh" and then "ah." But always at the end we see nature being threatened or destroyed by humankind and so we conclude "oh dear," here comes the heavy message. And certainly I feel that our stories, whether on the High Arctic, James Bay, Great Lakes, Sri Lanka or Madagascar, all end up the same way — ooh, ah, oh dear. Our challenge is to make shows that draw the viewers in, touch them emotionally and challenge them intellectually, so that afterwards, rather than giving in to gloom and despair, the audience is galvanized to act.

THE CBC — AN INSIDER'S VIEW

People often ask what it's like working for the CBC, and I think they usually expect me to begin levelling all kinds of criticisms. Like any large organization, the CBC has a lot for which to be criticized. But I am a diehard CBC loyalist and I believe Canadians get high value for their tax dollars spent on the CBC. Canada would be a radically different and poorer country without the CBC. This fact is reinforced every time I go to the United States and turn on the television. The overall quality of major U.S. network television is abominable, with the exception of PBS which only attracts minuscule audiences by comparison. CBC programs must compete with the major U.S. networks for audience numbers — and they succeed.

I am disgusted by those who suggest the CBC should be "privatized." It is because the CBC is a public network with a mandate to mirror Canada that it is the major vehicle for providing Canadians with glimpses of this country that create our self-image. Blind worship of the private sector will ensure that depth will give way to superficiality; ideas to dazzling visual gimmicks; and provocative challenge to blandness and inoffensiveness — because profit demands maximum audience with minimum discomfort.

Where the CBC must be faulted is in its dependence on commercial revenue. Not only do commercials eat into broadcast time ("The Nature of Things" loses over ten minutes of each hour to commercials), they can project a point of view that negates the program within which it occurs. How, for example, are we to respond to a hard-hitting environmental program if, cutting into it, are commercials proclaiming the corporate responsibility of the very industries being criticized? And when a sponsor is paying millions for commercials, can a network risk offending that business with critical documentary reporting? Of course not. Whoever pays calls the shots. CBC radio is more exciting and innovative than CBC television, in large part because radio has kicked its dependence on commercial revenue.

Many people in the regions outside Toronto resent the fact that Toronto is the hub of all English network programming. But virtually all countries have a single major centre for broadcasting. In England, it's London; in France, it's Paris; in Japan, it's Tokyo. The United States has New York and Los Angeles, but we, with one-tenth the people, try to have several centres, as well as an English and French network. It's far more than we can afford or fulfill. I don't think we have the resources to spread the CBC in many centres across the country. But having said this, I admit there is a tendency for broadcasters in the Toronto-Ottawa-Montreal triangle to think of that as Canada. My wife feels completely left out in Vancouver, whenever the host of a national CBC radio program announces something like, "Well, it's a lovely day today." Yes it is — in Toronto.

In Toronto, the CBC building I work in is far from the public's ideas of glamour. It's an aging, dumpy structure that is crammed with tiny jerry-rigged cubicles and rooms with poor ventilation. It is scandalous that no central facility for the CBC exists in Toronto. The corporation is spread out across the city in dozens of buildings, many old and dilapidated. It's inefficient and wasteful. On the other hand, I like being able to work within the Science Unit with a core of about twenty people. We are relatively self-contained and buffered from the rest of the CBC's bureaucracy. When I go into the office, it's like going into the familiarity and excitement of my research lab. There are always people to talk to about a recent shoot or research trip, to show off a partially edited film or just to discuss an idea for a story.

Journalists often claim that objectivity is the mark of the professional; I believe this is absolute nonsense. The reason we fight for Canadian newspapers, magazines and television is that our perceptions and priorities are different from reporters in the United States. We see reality through the special lenses of our personal experiences and background, and our stories can't avoid reflecting that. Men do see things differently from women, the educated from the uneducated, the rich from the poor and the white from the non-white. The only objective reporter is a dead one, so the idea that all reporting should be "balanced" and "impartial" is not achievable.

Just as we in the lab couldn't escape the administrative hierarchy above us, it is even more so at the CBC. During the short-lived Clark government in 1979, the CBC found itself being attacked for its perceived left wing, anti-Tory bias by the Conservatives. Consequently, CBC executives became very nervous about provoking controversy with the Clark government.

One of Clark's explicit aims was the sale of Canada's only major national oil company to the private sector. He wanted to dismantle Petrocan, one of the showpiece accomplishments of the Trudeau government's plan to gain some measure of control over national energy. I can understand the importance of Clark's proposal as a political symbol of his government's commitment to free enterprise and the ultimate good of marketplace forces. But I believe that people, through government, must have control of their own natural resources, so that public interest, not profit, sets policy. When Ed Broadbent and the NDP announced their plans to make the preservation of Petrocan an important issue, I was in complete and strong agreement with this position.

So when I was asked to write a letter to solicit support and contributions for Broadbent's Petrocan initiative, I was happy to do so. It was my understanding that the letter would be sent to all members and political supporters of the NDP, and it was. But I got into hot water with the CBC when, without my knowledge, my letter was published in the magazine *Harrowsmith.* (It also got the publishers of the magazine in trouble with people in the oilpatch who complained about the blatantly political nature of the letter.) I wasn't very concerned at the time, but I should have been.

My *Harrowsmith* letter somehow came to the attention of the upper echelon of the CBC. I had moved back to Vancouver for the Christmas holidays and I first became aware that something was brewing when Jim Murray called me. He was extremely distressed and told me that CBC management wanted me pulled off the air for the duration of the election campaign. I was floored. My immediate reaction was that my personal political activity had nothing to do with my television persona. It was common knowledge that CBC performers like Gordon Sinclair and Bruno Gerussi publicly supported political parties. But it was countered that they are entertainers, while I, as host of a program producing documentary reports, had to be objective. By writing that letter for the NDP, I had compromised my credibility. (At the same time, the CBC brass had no idea that Tory Minister of State for Science and Technology, Heward Grafftey, had been putting pressure on me to take a high government post.)

I've always been loyal and supportive of the CBC, but I wasn't prepared to be muzzled on the issues that I felt were important. Throughout my media career, I had made no pretense of hiding my beliefs and politics. For that matter, being on national television with shoulder-length hair, granny glasses and a headband, as I was on "Suzuki on Science," was a political statement. In the on-camera pieces which I write for "The Nature of Things," I state *my* views on the current story. I thought I was valued by the CBC *because* I had opinions and a point of view. Otherwise an actor could host the show much better than I.

I was further angered that the decision to take me off the air was made without consulting or confronting me. As during my crisis over the sabbatical leave at UBC, my immediate inclination was to quit in indignation, but I called Pierre Berton to ask for advice. "Tell them to go to hell!" he thundered. "They have no right to do it. They're making things up as they go. Tell them you refuse to get off the air, and if they take you off, you'll sue!"

I had always admired Berton, but his response endeared him all the more to me and gave me the determination to fight it through.

Fighting the CBC, however, is difficult. For one thing, you don't know who is sending out the edicts; you only know the people who deliver them. I wondered at the time whether this was just an excuse to get rid of me. And I worried about whether this affair would adversely affect "The Nature of Things." Did people fear that I was going to start soliciting for the NDP on air? Jim was caught in the middle. On the one hand, he had to keep me from a direct personal confrontation with the brass because he knew I might blow up and quit television altogether. On the other hand, he had to contend with his superiors who were quaking at the thought of catching hell from the Tories for giving a known socialist air time on a national program.

One of the arguments made to me was that, through the CBC, I was able to express my views to a large audience, and that brought with it a reciprocal responsibility to be "objective." As well, it was said, I was beholden to the CBC; I now had to consider the corporation when I acted in public. In the end, I did accept management's edict and left the show until after the election. The rationale that I reluctantly accepted was the following: I had publicly supported the NDP and its policy on Petrocan. Suppose in the middle of "The Nature of Things," the NDP bought air time and announced, "As David Suzuki has said, Petrocan ought to be retained by the public sector." This was about as likely as the NDP forming the next government and I knew that a simple phone call to the party would ensure that it didn't happen. But on the chance that political commercials in the middle of "The Nature of Things" might compromise the credibility of the show, I agreed to leave it for the duration of the election. Fortunately, we were able to fill those time slots with reruns of a series of special one-hour anthropological stories that had been done in the past. The CBC press release stated that I had left the show in order to take a more active role in the election, but it left a sour taste in my mouth that still persists today.

The CBC has the same problem that most large North American organizations have. It is built like a pyramid in which people higher up in the hierarchy believe they are more important than the people below. As people rise through the ranks, they inevitably acquire greater influence and power. And that is too often accompanied by their inflated sense of importance. Nothing indicates this better than their posh offices located in buildings or even cities away from where the actual production takes place. If the primary function of the CBC is to get programs on air, any decision within the corporation ought to be made with a view to helping achieve that goal. By the time executives have gained high administrative posts, however, they frequently end up making decisions that make *administrative* operations easier or better, often to the detriment of the production of shows. At CBC we get a stream of new administrative orders which often retard the speed, efficiency or savings in our films. Administrators have to realize that programs on air are all the public sees and evaluates. And production staff need to feel that management cares about their shows.

It is striking to compare large Japanese and American corporations, such as auto companies. Whereas Japanese workers turn in thousands of suggestions each year, their U.S. counterparts will make a few dozen. Most of the American suggestions are ignored or rejected: the Japanese act on over 90 percent of theirs. It's an indication of a very different worker-boss attitude. In Japan, everyone from the janitor to the CEO is interested in the company's output. In North America, the worker is made to feel removed from both the decision-making and the end results.

In North America there is a strong adversarial relationship between labour and management. It's almost impossible to instill a sense of pride and loyalty that can evoke an extra effort when needed. When I ran my lab, I kept my office door open; the fact that I was there working as long and hard as the others, affected the atmosphere and productivity of the group.

There has been relatively little public response to the severe budget cuts administered to the CBC over the past few years.

In part, it's because the CBC has made a tactical error. Instead of cutting whole services — for example, all morning programming or overseas broadcasts — the CBC has chosen to spread the cuts around. So there are no priority program areas that are protected, but only a democratic sharing in the attrition as a way of coping with the decline in per-show dollars. The implication is that there must be a lot of fat because the CBC seems to be going about with business as usual, in spite of the cuts. But there is a perceptible decay if you look for it. "The Nature of Things" has chosen to cut back on the number of original shows in order to maintain the program's quality. But that means we run more purchased programs and reruns. Other series have absorbed the reduced budget by cutting corners on production costs and living with a gradual decline in quality. Either way, the viewing audience loses out.

Clearly, the absence of a public outcry at government-imposed cuts to the CBC reflects the failure of the corporation itself to educate Canadians about public broadcasting. The public audience for the CBC is there — the numbers show it. But the viewers haven't understood how the CBC differs from private networks and how its programs are unique.

I believe it is a mistake for CBC executives to be the only public spokespeople for the corporation. They may have clout in government offices, but the public doesn't know them. Unless an executive has the public profile of a Lee Iacocca, it is the "stars" whom the public recognizes. It takes millions of dollars to transform a Vicki Gabereau, Knowlton Nash or Hannah Gartner into a celebrity, and it's a shame to waste that recognition. I think the CBC should have an ongoing campaign in which it makes its stars accessible to the public in different parts of the country. There should be an opportunity for people to meet the CBC's celebrities in a more intimate way and hear what their programs, the CBC and the public mean to them. Unfortunately, when I proposed this several years ago, the powers-that-be rejected it as too self-serving. It may be self-serving, but if we believe in what the CBC represents we have an obligation to explain that to the people.

THE OTHER MEDIUM — PRINT

While most of my time as a science popularizer has been spent in the electronic media, as a scientist I still have a deep-seated respect for print. Television, radio, print — each medium has its virtues and limitations. Television's power is the picture and its ability to evoke emotional responses. Television can give visual perspectives that cannot be matched by sound or print. Pictures are very effective, but they are also limited. Abstract ideas are difficult to convey with pictures. Radio, on the other hand, uses every listener's brain to create the images. "Talking heads" are radio's strength, and sound effects can trigger a multitude of impressions in a listener's imagination. But radio and television are ephemeral — they are broadcast and gone. Though recording devices now let us capture programs, more effort is required to record and replay than to read printed material.

Print is tangible — it can be touched and hefted. It is permanent. The reader can proceed at his or her own pace, re-read or jump. As Robert Fulford once told me, if print had been invented *after* radio and television, it would be regarded as a wildly innovative medium — compact, portable, requiring only a brain for use. In the electronic media, because images change rapidly, we can, on occasion, "slip things past" the audience. When we don't know a precise fact or number, we can "waffle" and let it go. But with print, the author is exposed — it's all there in black and white. I may wince or twitch when I see myself making a mistake on TV, but the moment passes quickly. But when I see an error in my writing, it is there to stay. Writing for print is very hard work and I am filled with admiration and envy for people who can crank out columns or reports daily.

My venture into print journalism began during the seventies for a magazine called *Science Forum*. This was primarily a vehicle for communication between people in science, industry and government. It came out six times a year and never had more than a couple of thousand subscribers. In 1976, I was asked to contribute a regular column to *Science Forum* and was happy to do so. But soon it became a burden. The magazine's editor,

Wayne Campbell, had to hound me to meet deadlines. Since I wrote the articles for free, Wayne had no leverage to make me fulfill my responsibility. More than once I didn't get a column in. I always protested I was too busy, but in fact, I didn't have much to say. I was relieved when the magazine folded.

In those days, I hadn't yet developed a philosophical framework that would give coherence to my ideas and hone my arguments. I was still trying to demystify science with a scattergun approach — exploring its marvels, pointing out its hazards and raising moral and ethical issues helter-skelter. So in attempting to write a column, I often found that my ideas didn't "go anywhere." As well, my expertise still rested heavily on genetics — I hadn't branched out, as a broadcaster, to see genetics as a part of the technological explosion taking place. Working on "The Nature of Things" and then "A Planet for the Taking" took me into a whole new realm of ideas and perspectives which I wanted to express and share.

In 1984, when the newly elected Conservative government announced severe cuts to the CBC budget, I was so furious I wrote a long article defending the corporation. It was over three thousand words long, and I went through at least eight revisions, printing it out each time. Had I done it by typing or writing, it would have taken days; with a computer, I was able to do it in hours. I submitted the article to the *Globe and Mail,* which turned it down flat. I then sent it to *The Toronto Star*, which immediately accepted it and printed the piece in its entirety. That got me thinking more seriously about expressing my ideas in print.

In 1985, I began to write a regular column for *The Toronto Star.* After having failed to meet deadlines for *Science Forum,* which only came out six times a year, I was nervous about meeting a weekly commitment. But I did it, on time every week for fifty-two weeks. On radio and television, my programs are reports on specific topics. I seldom have a chance to inject my own opinions. Newspaper columns allow me to be a gadfly and express my personal opinions on a broad range of subjects. I enjoy writing them immensely.

My columns for *The Toronto Star* only appeared reliably in the Toronto area, and often when picked up by other newspapers, they were severely edited and changed. It was frustrating to have a localized audience and to lose control of my writing when the articles were syndicated. So after a year I quit writing the columns and took some time to think about it. I later approached the *Globe and Mail* because it has national distribution. On January 1, 1987, I started writing again, this time a weekly piece for the *Globe and Mail.* So far, so good.

TRICKS OF THE TRADE

It is remarkable how easy it is to grasp the tricks of film and to "see" the story created out of illusion. In real life, we don't have a zoom lens on our eyeballs, an ability to cut from one perspective to another instantly, or to change locations or time period. Yet that is what is done with film, and we understand the intent immediately. I watch in amazement as Sarika, my three-year-old daughter, grasps the essence of the story through the montage of shots assembled. I often wonder what this does to her sense of reality.

It is instructive to know a few of the "tricks of the trade" because it helps to understand that movies are creations, not reflections of reality. When I first began to work with the Science Unit in Toronto, I realized what an illusion film creates. I watched the editing process as one sequence was being put together. It was in a report on the physical condition of children in Saskatchewan and the sequence involved a girl sitting upright on a machine, then bending forward to touch her feet and straightening up again. It required at least four different shots — a "close-up" on her face, a "wide shot" as she began to bend forward, a shot of her face again from the side, and then another wide shot as she returned to the sitting position. When edited together into a ten-second sequence, these gave the viewer a sense of seeing one continuous, smooth motion. But it was actually our brains that had interpreted those shots that way. In fact, the camera or the lens had to be separately positioned for each of those shots, while the girl repeated the exercise several times to give us plenty of material to edit together.

Have you ever watched those interviews on "The Fifth Estate" or "60 Minutes" where an inquisitor is grilling a subject and the camera cuts back and forth from interviewer to the interviewee? Unless it's a pretty dicey situation, where the interviewer is suddenly going to accuse the interviewee of some unspeakable act, the entire thing is shot on *one* camera. Two cameras mean a lot more equipment and crew, and that means money. So, generally, the subject's face for the entire interview is shot over the interviewer's shoulder. Someone carefully writes down all of the questions the interviewer asked (they're called "re-asks"). Usually a wide shot will be taken that shows both people at once, so that there is some indication that they were actually in the same place together. The camera then is repositioned to film the interviewer's face and he "re-asks" all of the original questions. Often he may be looking at a blank wall or one of the crew because the interviewee is no longer there.

I remember doing this in 1982 when Wayne Gretzky was in a hectic race for a new scoring record and everyone wanted to interview him. I was producing a report on how body movement is controlled. When you begin to measure the length of time it takes for impulses to travel along neurons, it becomes clear that the movements someone like Gretzky makes don't make sense in neurophysiological terms. The time interval between his *seeing* an opening and his acting by deliberately commanding muscles to respond so that his entire body moves to make a shot, is too short for electrical impulses to travel down nerves from the eye to the brain and out to the muscles. So we wanted to ask Wayne what goes on in his head when he makes a spectacular shot or move.

It was quite a job to get through all of the red tape, but eventually we got a PR person from the Edmonton Oilers to give us permission to film Gretzky at a practice in Vancouver. I had the crew in the stands all set up as the players came off the ice. The trouble was that no one had bothered to tell Glen Sather, the coach, who was trying to protect his superstar from the likes of me. When I went up to Sather, he waved me off and said rudely, "No interviews for anybody on the day of the game." I was shattered. We thought we had it all arranged.

But Sather is a mischievous guy. He allowed us time to stew, and when I didn't protest, he finally skated over grinning and said, "But I like the stuff you do. You've got five minutes." Now to a camera crew, five minutes is like being offered five seconds, but Gretzky was a superstar. So we got ready. Gretzky came out and, with no chitchat, we put the microphone on him, asked him the questions and were finished in less than four minutes. "Is that all?" he asked in disbelief, and raced off happily. In any case, long after Wayne was gone, I was earnestly "re-asking" questions to the empty stadium.

Another high-pressure interview was with Denton Cooley, the famed heart surgeon in Houston, who has performed more heart transplants than any other person in the world. He has been known to carry out over fifty heart operations in a day. It's an assembly line. He works in a series of operating rooms laid out like spokes on a wheel. Each room has a complement of surgeons, nurses and anaesthetists who prep the patient and open him or her up. Cooley simply passes from one room to the next picking up the chart and then performing the crucial part of the procedure. Not surprisingly, he reached fifty thousand open-heart surgeries long ago.

We had arranged to do an interview with him at the end of a work day. I watched him perform from an observation gallery above the ORs, and although he made it look effortless, it wasn't hard to imagine the stress this put him under. We had everything ready in his office when he arrived, drained and listless. As he did some paperwork, he ignored our presence in the room and his secretary informed us that we had twenty minutes to do the interview and have our equipment out of his office. Again, this put a lot of pressure on us. But like so many high-profile people, Cooley had done interviews so often, he was able to suddenly turn on some switch to give us what we wanted quickly, albeit rather dispassionately. We didn't dare do any re-asks that time.

Although we didn't have to use re-asks with Cooley, there is a great temptation to put the interviewer against a background somewhere else in Houston or even back in Toronto and shoot the re-asks. That's because a re-ask not only allows the audience to see the face of the interviewer, it gives an "editing point" where the film is actually cut or where a shot on tape ended. When an interview is shot, we usually have much more film than we can use. A transcript is made of the interview and then we do a "paper edit," selecting parts that say what we want. But if it's in a "static" shot where the camera lens doesn't change its position, we don't like to cut out parts of the interview because, when the parts being used are put together (this is called a "jump cut"), a person's face does an obvious shift from one position to another at the edit point. News reports often put in jump cuts, but we prefer to stick in a "cutaway," a shot of something relevant to what is being discussed (such as an open heart if it's being discussed) or the interviewer nodding in agreement, smiling, looking serious, etc. The latter insertion is called a "reaction shot" that is vital for the editor. It is often difficult for the interviewer to do these reactions convincingly, because they're filmed after the interview.

It's very time-consuming to get these shots and usually the scientist being interviewed has no idea what to expect when he or she agrees to do an interview. Often when we contact someone, he or she is flattered, excited, and quite keen to do the interview. But once the hours go by and little seems to be happening, the enthusiasm wanes and the resentment begins creeping in. We may totally disrupt the lab, yet the scientist knows that he may only end up as a couple of minutes in the show. Part of my job is to keep the scientist occupied, engaging in chit-chat to divert him from the time it's taking. Experience shows that a scientist always talks about his work best the first time, so I try to avoid discussing the material we want him to cover on camera.

On rare occasions, I run into moral dilemmas in my interviews. I once did a long interview for radio with William Shockley, the

Stanford Nobel Prize-winner who has publicized the view that blacks are genetically less endowed intellectually than whites. From the moment I arrived in his home, I tried to be pleasant and chatty, because I was a guest in his house and I wanted him to be relaxed for the interview. But I had to maintain strict control or I would have lost my cool and started to attack him for his outrageous claims. Shockley is a passionate, articulate man who also carries enormous clout because of his Nobel Prize. Never mind that he's an engineer, not a geneticist — people are impressed with what he says because he's a Nobel Laureate. Though people may be shocked by Shockley's statements, they always remember that some Nobel Prize-winner says blacks are genetically less intelligent than whites. Is it better not to have him and others like him on programs to reinforce latent prejudices? But that would be censorship. I haven't worked it out yet.

When I meet a scientist to be interviewed in the United States or abroad, I introduce myself as a broadcaster. Of course, if they ask about my background, I immediately inform them that I am a trained scientist. Usually they don't ask and I don't say. I'm not being deliberately deceptive, and on a few occasions, it has provided interesting insights into a scientist's attitudes. For example, once I went to the home of the great astronomer Fred Hoyle in Penryth, in the Lake District of Britain. He didn't know me and accepted me as a journalist. What amazed me was the warmth with which he received me. He showed me around his home, shared his time generously and never once indicated that he felt I was wasting his time.

Another time I attended the annual meeting of the British Association for the Advancement of Science. The president that year was a Nobel Laureate who had done his work on the physiology of muscle. But at that time, he had been making public statements about the inheritance of human behaviour and the apparent decline in "quality" of human beings. For me, this is like raising a red flag.

I began the interview with the president of the BAAS by asking for his views on human heredity. All he knew was that I was a

Canadian broadcaster, and he replied to my question just as he had been quoted in the press. I let him speak at considerable length and finally interjected by pointing out that one of his eminent colleagues, Walter Bodmer, the human geneticist at Oxford University, had a very different position. When I mentioned someone he knew who was an authority and a big name in genetics, his attitude immediately changed. He began to waffle, "Well, yes, what I'm saying is speculative of course, but . . ." I continued to press him to be specific. It was clear he was becoming more and more uncomfortable and wished I would end the interview. Finally, I told him, "You are a Nobel Prize-winner and that carries enormous prestige with the public. When *you* make a statement, it is taken far more seriously than one made by an ordinary scientist. You are making statements that have enormous social significance, yet they lie outside your area of expertise. Don't you think it's irresponsible for you to do so so lightly?" He had nothing to say and rushed away as soon as I finished. Had he known from the start that I was a geneticist, I'm sure he would have acted differently.

All too often, when we pull in to shoot an interview in a lab, the cameraman will say, "This doesn't look very good" or even "This doesn't look like a lab," and will begin to rearrange things so that it looks the way *he* thinks a lab should look. But that grungy, boring lab that first greeted us is the way labs are. The second year after "Quirks and Quarks" began, a very aggressive PR man for CBC wanted to launch a high-profile promotion of the radio show. So he arranged to shoot a "promo" to show on CBC television. I arrived to find a classic Hollywood impression of a lab. The table was covered with Florence and Erlenmeyer flasks, lots of curled tubing and coloured liquid in the containers. He had dry ice in the liquid to give off lots of foggy vapour. I was appalled. To top it off, his script called for me to talk about the dangers of science and ended with me pulling a child onto my lap on the punchline, "You can't afford to miss it." I was furious at this caricature of science and tried to get the script and scene changed. We argued, and when he wouldn't back down,

I simply stormed off the set and never worked with him again.

Finding a subject who is a good "talking head" is like discovering pure gold. We scour the land in search of someone who can present material simply, clearly and enthusiastically. When we find such a person, we tend to milk him over and over. You'll often see the same scientist being interviewed in many different shows. That's why someone like Carl Sagan becomes a media star. Not only is he intelligent, witty and attractive, he's a great performer on camera. We find that the most reliable on-camera performers are scientists from Britain. I believe this reflects their culture, in which debate and discussion are important not only in school but in pubs and homes. Americans are also often good interviews — they're relaxed and candid. The poorest as a group by far are Canadians. We are a modest people and apparently not a highly articulate group, at least as far as film goes.

Before an interview, researchers check out the scientist. If he's the only or the main person involved in the topic being reported, then we have to make the best of him regardless of how he performs as a media person. But occasionally in a prefilm interview, a scientist will be terrific — articulate, clear and often saying outlandish things over which we rub our hands in anticipation. But once seated in the lights, with all eyes on him as the camera starts up, he turns into a boring, incomprehensible lump. In his mind, he is now talking, not to the public but to his colleagues who will criticize him for any inaccuracies or exaggeration. Now he begins to qualify his statements, to use jargon and to avoid speculation.

Once we drove for several hours to a remote location to do an interview with an expert on the ocean. He was terrific. We met him at a motel and he was talkative, funny and said wonderfully provocative things that we love to have in shows. When we set up the camera and started to roll, we were shocked to see him transformed into a dull, pedantic academic. "Cut!" yelled the producer after a minute. As soon as the camera stopped, the expert was his old self, joking with the crew and saying outrageous things. He was completely unaware of his personality change on camera, or perhaps he believed that it was proper to be more sober and serious to the "public." The interview was a disaster.

Rarely does a subject completely freeze on camera, but when that happens, it's very difficult to thaw him. Indeed, I believe you can tell within the first minute of an interview whether it will be any good. Occasionally an older scientist may start off strongly, but will gradually fade as the camera grinds on. And in very rare instances, a rather ordinary interview may suddenly come alive.

In August 1986, I taped an interview in California with Jonas Salk, the famous developer of the polio vaccine. He's an extremely busy man and it took me months of pestering him before he finally agreed to an interview. He emphasized that he would only give me half an hour. I arrived half an hour early to set up my gear and then he kept me waiting another fifteen minutes beyond the appointed time. Finally, he invited me into his office and began to listlessly answer my questions. He had obviously answered the same questions on work he'd done long ago countless times. I got what I needed but was disappointed by the absence of passion in his statements. As I packed up my gear, I innocently asked him to explain the difference between his vaccine (which contains dead viruses) and the Sabin vaccine (which has a crippled live virus).

Suddenly Salk came alive. I could almost see smoke flowing from his nose. I turned my tape recorder back on as Salk launched into a spirited demonstration of the superiority of his vaccine. I had touched an issue that has become his life's mission — to prove the superiority of his vaccine over Sabin's. Salk was clearly hurt and bitter that the Sabin vaccine had displaced his own, and he certainly convinced me that the Salk vaccine is superior. He went on for another forty-five minutes and I finally had to terminate the interview because I had another appointment.

While an interview is going on, the producer and I are listening carefully to hear whether the subject delivers something succinctly and clearly that we'll be able to use. So we're editing the interview in our minds as we go. And we may rephrase a question or go back over the material in a different way, trying to get what *we* want. Sometimes it's very frustrating because I know exactly what he should say and how he should say it, but *he's* the expert. I often feel guilty when someone has totally bombed out and we already know in advance that we will not use the interview. But we have still taken his time and accepted his hospitality. As we extricate ourselves, everyone tries to be polite and enthusiastic, thanking him profusely and sometimes saying, "Terrific! That was great. Thanks a lot!" knowing it cannot be used.

Once I did an interview for radio with a novice who had never been interviewed before. He was nervous and asked whether we could edit out something if he made a mistake. At this point, the producer who was trying to help him relax, told him about an interview we did with a famous economist who had a bad stutter. Imitating the economist's stammer, she told the scientist how "we were able to edit out all of the sssssttttutteriiiiing and make him sound good." Once we started the tape and began the interview, the scientist, who had spoken so clearly during our pre-interview conversation, developed a terrible stutter! The producer and I carried on with the interview, avoiding each other's eyes the whole time. The interview seemed interminable and I couldn't wait to end it. Once we got out of the lab, the producer and I practically screamed at our stupidity and insensitivity. Fortunately, the scientist never held it against us and did interviews with us again.

STARDOM

The power of what it means to be a star came home to me when I was working in Toronto while Tara was living in our home in Vancouver. She called me one night in a state of excitement and informed me that Charlton Heston was making a film in Vancouver and that his company had discovered our house and wanted to use it for one of the scenes. One of Heston's associates had called on her to ask to rent our house for three days. I immediately discouraged her. Even our small crews of four or five can be quite disruptive in a home. More than once I've seen what a hot light can do to a wall or ceiling. "Tell him to forget it," I told Tara firmly, so she reluctantly agreed.

The film was called *Motherlode* and had been written by Heston's son, Fraser, whose wife is from Vancouver. In fact, they were living in West Vancouver. The next night Tara called again. "I've agreed to let them film," she told me. "Fraser Heston came over and he's very nice. He assured me they'd be very careful, so I agreed." I argued vehemently against it, describing the chaos and damage that would ensue and managed to convince her once again to turn it down.

The next night, Tara called again. "Charlton Heston came to see me!" she exclaimed, "and he's *going* to film in our place!" There was no arguing — when Moses/Michelangelo comes over personally, it's pretty hard to say no. That's the power of stardom.

In a smaller way, I also experienced the weight of celebrity. We were shooting a film about what environmental effects drilling for oil in Lancaster Sound in the Arctic would have. We landed in Resolute Bay, and after dumping my stuff in the hotel, I took a drive into the Inuit part of town. As I got out of the car to look around, kids came pouring out of the houses, staring at me and calling my name as if I was a Hollywood star. I learned later that Resolute had a satellite dish and that the kids were avid watchers of "Science Magazine" and "The Nature of Things" because they thought I was an Inuk. I felt honoured.

A major plus of my work in broadcasting has been the famous people I've met. I've interviewed a lot of people whom I've always admired, like Margaret Mead, Paul Ehrlich, Stephen Gould and Wayne Gretzky. Meeting Germaine Greer was particularly memorable. She was in North America promoting her book, *Sex and Destiny*, and we wanted to film her for "A Planet for the Taking." She had a gruelling schedule but we managed to make an appointment to interview her in Vancouver. We were told we had no more than one hour. I was pretty nervous about meeting the notorious author of *The Female Eunuch*, but she turned out to be charming. We interviewed her in my house and she and

Tara got on very well. After we shot the interview, she stayed on for a couple more hours talking about everything from genetics to children to English gardens.

Before Carl Sagan was a household name, I invited him to come to Vancouver to be on a series I did for the provincial educational channel. I have always been an admirer of his television presence. On talk shows Sagan is without peer; he is so knowledgeable and articulate. After we had finished the show, I drove Carl out to the airport and got so involved in our conversation that I missed my turnoff. So I did a U-turn at the next corner. Immediately a police car appeared and pulled me over. Cursing, I jumped out and ran up to the police car to try to talk my way out of a ticket. I didn't succeed, but when I got back in the car, Sagan was sitting there speechless. He told me he had been horrified when I got out and went up to the police car. "In America," he informed me, "you'd probably have been shot."

Tara and I received an award for a book we were writing together and it was handed to us at a special ceremony by Governor-General Jules Léger. We were astonished to see Margaret Trudeau seated in the small audience. During the reception afterward, Tara and I chatted with her, pleased that she was there. "Do you have to fill in for your husband at things like this?" I asked. "I don't *have* to do anything," she retorted. She informed us that she had come out of interest and to support a fellow British Columbian.

Later, when I received the Order of Canada, the public knew that Pierre and Margaret were having marital problems. Trudeau was at the reception because David Lewis was one of the awardees and a lot of politicians had come to honour him. Tara and I were sitting with Karen Kain, another recipient that day, and her parents when Trudeau came over to chat. I babbled out at one point that when we had met Margaret several months earlier she had been very interested in Tara's difficult decision to go on for a Ph.D. in Wisconsin. Margaret had recently gone south too, to New York, also to further her career. Trudeau immediately feigned anger at Tara. "So *you're* the one who's been giving Margaret those ideas!"

When people suddenly encounter someone they recognize from TV, they are often flustered and call out the wrong name. Roy Bonisteel, the host of "Man Alive", told me he was once crossing the tarmac at the Edmonton Airport when a fan breathlessly accosted him as Peter Gzowski. If you know both of them, you also know that's some mix-up. Gwynn Dyer, the host and writer of the brilliant series "War," told me what happened after his series ran on PBS in the United States. He was waiting in a line at Kennedy Airport in New York when he noticed someone staring at him. Finally the person came over and said, "You're a Canadian, aren't you?" "Yes," answered Dyer. "You had that series on PBS, didn't you?" "Right," said Dyer. "You're David Suzuki!" gushed the fan. I'm honoured by that story.

I had a similar encounter in a Chinese restaurant in Toronto. I was eating alone when one of the Chinese waiters did a classic double-take. He walked by me, glanced my way, then turned right around and stared. "Hey, you're that guy on CBC television. Right?" he exclaimed. "That's right," I replied. "I know you," he announced to the room. "You're Mel Tsuji!" He got the right ethnic group, and Mel is a reporter for CBLT in Toronto. So I guess we even look alike to our fellow Asians.

In reflecting on these stories, I'm struck by the incredible impact of the media on people's lives. The media have bestowed upon me an awesome privilege and responsibility, and opportunities to share new ideas with a wide audience and to meet memorable individuals. Broadcasting has also been a demanding, frustrating, frenetic way of life, but I wouldn't give it back for anything.

The family matures (l to r): *Back row:* Aiko, me, Marcia. *Front row:* Dad, Dawn, Mom.

10

11

On location in southern Indian temple for "A Planet for the Taking."

Completion of the China shoot in Shanghai after five and a half gruelling weeks. (l to r): Neville Ottey (camera assistant), Jim Murray (executive producer), Archie Kay (lighting), Richard Longley (researcher), John Crawford, (sound).

12

13

This is the fabled Hunza Valley, high up in the Himalayan Mountains where people are reputed to live well into the hundreds. That's Heather Cook, one of the producers of "A Planet for the Taking."

The Yoruba of Nigeria have the highest incidence of twins in the world.

(right) This immense tree on the Queen Charlotte Islands may be a thousand years old.

(below left) The pyramids of Gizeh for "A Planet for the Taking."

(below right) Sea urchins are the main food source of the sea otters reintroduced here at the Bunsby Islands off Vancouver Island.

14

15

EPILOGUE

REFLECTIONS ON REACHING FIFTY

FIFTY IS A WATERSHED YEAR in one's life. I reached the half-century mark last year as I began to think about this book. I sometimes still feel like an immature college student, but now life no longer stretches ahead endlessly. There are also the visible changes of ageing that no amount of exercise and dieting can reverse. It's a time to take stock, yet now more than ever, there seem to be so many more things to do.

The 1980s find me still in the process of metamorphosis brought on by a new family, Tara's professional career and my renewed focus on children. But it's all too immediate for everything to have been sifted into a coherent picture yet. So here are a few snapshots punctuated by recent essays I've written.

A NEW FAMILY

Tara and I had agreed before we were married that we would like to have a family together, but there was no urgency. We were wonderfully free — we travelled together to the United States, across Canada and to Europe. And when Tara decided to go to graduate school in Wisconsin, there were only the two of us to worry about. We could move easily between Vancouver and Toronto.

In early 1979, Heather Cook produced a program on infertility for "The Nature of Things." The most striking information in the film to me was the steady decline in fertility of women from their early twenties on and the more rapid loss of fertility after thirty. Tara had just reached thirty, and after I saw the film, I suggested if we were to have a child, we'd better start right away. So our first daughter Severn arrived, on November 30, 1979, and I was gifted with a second chance at being a father. She was named Severn after the valley where Tara's dad came from and Setsu after my mother. I have tried not to miss out on the experiences of parenthood, even though my life still involves a lot of travel. It is always a challenge.

* * *

ON THE ROAD WITH BABY

(This article first appeared in *Vancouver Magazine,* August 1981.)

As an only son raised with three sisters in a traditional Japanese family, marriage to a committed feminist forced my ideas on the role of women to change. True, I had been long involved with the issue of minority group civil rights and the obvious proposition that equal pay for equal work, freedom from harassment and equal opportunity be guaranteed for everyone, but it is quite another thing to know in a gut way the enormous inequities of sexual roles in society today. Nothing brought this home more intensely than my experience with my daughter, Severn.

As a father of three children in the 1960s, I delighted in the time-honoured role of "head of the family," pursuing professional goals, providing a home for the family, coming home to play with the kids, give them their baths and read them to sleep. Raising children was a snap. All of them started camping before they were a year old, and, of course, I did "dad's jobs" of putting up the tent, catching and cleaning the fish and playing with the kids. Lots of fun.

With the children grown up (Tamiko is twenty-one, Troy nineteen, Laura seventeen), and sixteen years after having a new baby in the house, I was blessed by the start of another family and the opportunity to relive those years with new eyes. How the times and I have changed! On the one hand, the participating hospital experience before, during and after Severn's birth was the reverse of my earlier banishment to the waiting room, and certainly in the past I never found myself with four male friends, all of us carrying our babies and arguing about the optimum time to stop breast-feeding.

Over the years, my interests in science have shifted to the study of how genes control development, and being less driven to academic "success" than I was as a budding assistant professor, I have been able to watch Severn's growth, coordination and communication with new-found awe. The rate of change in those early months is absolutely astonishing, but, for me, also frustrating. As a broadcaster who must travel often, I have been distressed to miss even a day of Severn's progress as she seems to change daily. So, when she was eight months old and my wife Tara had stopped breast-feeding, I decided to take Severn *with* me on the next week-long trip to Toronto. Feminist Tara was in no position to deny this, and there was just a slight hesitation before she nodded assent.

Not liking disposable diapers, I carried cloth ones, although I now know that disposables are made for use on planes and that barf bags are great for keeping dirty diapers in. Anyway, I set off with a small shoulder bag of my things, two huge suitcases stuffed with baby toys, clothes, etc., plus a carry-on suitcase of articles for baby during the flight, a stroller, and Severn strapped to my chest in a Snuggly.

Few things can be more frustrating than trying to appear relaxed and in control while every hand and armpit is stuffed with baggage and a crying baby. But everyone was extremely helpful (a clerk pointed out that seat 22H on a 747 is ideal, being on the aisle with no seat in front and a good view of the screen). Of course, pre-boarding is great, and I got settled with an empty seat next to me. Severn was terrific, happily taking a bottle during takeoff and landing to clear her ears as the pressure changed. She slept during most of the trip in the seat next to me, and played happily with the toys Air Canada has for kids. The only crisis I had was having to go to the toilet while Severn was sleeping. Too embarrassed to ask my neighbour, a male, for aid until an imminent bursting bladder overcame that reticence, I then found him pleased to watch that she didn't wake up and roll off. Since that experience, I have offered to hold babies for mothers travelling alone, and know exactly how they feel when they rush off.

Overall, the trip to Toronto was never more of a pleasure. We arrived late in the afternoon on Sunday, and I had all of Severn's things unpacked in the apartment an hour later. Tara had prepared enough baby food to last a day, so Severn and I rushed out to stock up on a week's worth of vegetables and meat. When Severn became tired, I pushed the bed against the wall so that she could sleep between it and me. When I called Tara later, I was elated and rather cocky.

Monday morning I woke up to the alarm, still fuzzy-headed from the three-hour time difference. I didn't want to wake Severn, but how could I take a shower and fix her breakfast without worrying about her falling off the bed? After piling up pillows and suitcases to keep her wedged in, I ran to the toilet, roared through a shower and then made up pablum and a bottle of formula, meanwhile checking the bedroom every few minutes. Finally, I woke her up, changed and dressed her, and played with her during breakfast. One thing I learned right away: if you are about to go on camera, never wear your studio clothes *before* feeding a baby. They have an uncanny ability to sneeze or spit pablum onto a shirt front.

Our arrival at the office at 790 Bay created a minor sensation. It is truly wonderful how women respond to a baby, and I was soon surrounded by people wanting to hold Severn. The editor was terrific, letting Severn crawl around on the floor, safely within a "kiddie korral." She was happy to play, to drink milk and sleep on a blanket on the floor. For lunch, I would order fish and a vegetable, then, as discreetly as possible, chew up small pieces which I would then stuff into Severn's mouth. This may sound revolting, but it beats carrying a blender; and anyway, who finds kissing disgusting?

Like many Japanese males, I never learned to cook. So after work, I would hurry back to the apartment, boil up broccoli and carrots, cook a bit of meat and blender the whole rudimentary mix for Severn. Friends' invitations were not accepted that week as I found just caring for baby to be a full-time job that kept me busy enough to eat only her leftovers and lose five pounds! After dinner, there were dishes to clean, usually the floor to scrub, clothes and diapers to wash, a bed to make, and lots of playing to be done. It is incredible how carefully a crawling child must be watched; a stray thumbtack, paperclip or pencil unerringly ends up in her mouth.

During some midweek filming, the camera crew proved to be extremely considerate and generous, the lighting man taking Severn and keeping her occupied when I had to perform. In fact, in one of the show's segments, Severn may be seen in the background, playing in a car. I was feeling pretty pleased with myself until one of the producers pointed out that had *she* ever tried to bring her baby to work, she would have been sent packing. Males with babies are novelties. People thought it all "cute" and besides, as host of the show, I could "get away" with a lot. That crashed me back to reality.

The days sped by, each amazing me with the number of times a child needs attention — not just to be fed, changed or put to sleep, but because of boredom, a call to share something interesting, or just a need for cuddling. Simple things became enormously complex. Taking a bath, for example. The maneuvers would need careful planning: laying out all of baby's things and a large towel next to the tub. In the bath, no problem, but then baby has to be dried off, diapered and left to play while dad does a very quick once-over before leaping out to dress himself and Severn.

The secret to it all turned out to be *speed* — running and moving like a flash to make sure the baby was almost never out of sight where she might hurt herself. Her naps became times when I would tear around doing the housework, praying that she would sleep another fifteen minutes. Each night, with one belt of scotch, I would drop off to sleep by 10 o'clock.

The intimacy of our contact was something entirely new. I was with her twenty-four hours a day for seven days, the most sublime moments of which were waking in the morning to find her laughing and gurgling six inches from my face. When she grows up, she will remember none of it, but for me it is an indelible memory that fills me with delight.

Returning to Vancouver was night and day compared

to the flight out. The aircraft was packed for one thing, although I was happy to discover that tables in the toilets facilitate changing babies. Severn had a high fever, and halfway home threw up all over the Snuggly and my shirt. Of course, I had no change of clothes with me, and cleaning up as best as I could still left me acutely conscious of that distinctive vomit smell emanating from my chest. I understood, as I worried that baby would disturb the people around, why Tara had felt so self-conscious nursing Severn in public. Any whimper set off waves of new concern and potential embarrassment.

Back in Vancouver, Tara seemed like a new person; she had been to plays and visited with friends every evening while taking care of numerous jobs during the day. She had not worried (which was a great compliment) nor missed us at all, and was bubbling with excitement and happiness at our return. Her rejuvenation alone would have made the whole thing worthwhile.

Severn has since accompanied me on other trips, all of them delights although none with quite the impact of that first. It was illuminating in other ways too. I led a simple, monastic week with Severn, but many mothers with small babies must also work, cook, shop, entertain and handle all of life's routines. Nothing has done more to show me the enormous physical and emotional demands a woman operates under; I strongly urge every father to take care of an infant for a week to prove it for himself. And for the joy of it.

* * *

By the time Tara was pregnant for the second time, I was well into filming for "A Planet for the Taking." Two crews were filming in different parts of the world and I shuttled between them. At one point, I flew from Japan straight to the Kalahari Desert in Botswana.

Our second baby was due April 1, 1983, so we had scheduled two weeks before April 1 and two weeks after so I could stay in Vancouver with Tara. Then I was slated to meet the crew near the southern tip of India. April 1 came and went with no sign of the baby. The days ticked by, and as we approached the 15th, we began to panic. I sent telexes to the crew to let them know I might be delayed, and they received garbled messages that seemed to suggest some kind of tragedy. I even gave Tara a folk remedy of castor oil that is reputed to induce labour, to no avail. Finally, Sarika Freda (after Tara's mother) was born on April 15, and I was only able to stay long enough to get Tara and the new baby home before taking off for Madurai.

Madurai is a large city, but the phone system is primitive and overseas calls are impossible to make. Tara and I lost contact for weeks. I went on to Delhi in northern India, then through Karachi, Pakistan, to the ancient ruins of Mohenjo-Daro, one of the earliest known human settlements on the Indus River. From there, I flew to Lahore and on up the Hunza Valley, deep in the Himalayas. Finally, I caught up with another crew in Greece. From there I went on to London, England where Tara had come with Severn and Sarika. It had been almost a month since I'd seen my new daughter.

"A Planet for the Taking" was the most severe test of our marriage. We were separated for weeks at a time, and Tara was left managing the household with two young babies and worrying about two sets of ageing parents. Her father had developed angina and we were concerned about his heart. My mother had steadily lost her memory from Alzheimer's syndrome. Indeed, on the day after Sarika was born, when my parents came to the hospital to visit, my mother had wandered off and disappeared. What had started as a joyous event turned into a terrible panic as we first searched the hospital, then scoured the city for my mother. She was found at three o'clock the next morning.

Nor was my life a picnic; I was working at a gruelling pace. I would fly into a location, learn my lines, deliver them on camera, then take off for the next place. There was no time for sightseeing or rest. In one ten-day period, I flew from London, England to Paris, drove to southern France and back to Paris to fly to Rome, then on to Palermo to Cairo to Tel Aviv to Munich. At each place, I did on-camera pieces. I remember it all as a haze and only appreciated what we had done back in the screening room in Toronto where we looked at the "rushes."

We have survived "A Planet for the Taking," but our lives have not become simpler. Tara applied for a faculty position at Harvard University in Cambridge, Massachusetts and was offered the job. It was a wonderful recognition of her academic record and abilities, and we were euphoric after she got the job. But with two young girls and two parents working in different cities, our family has become a modern quest to fulfill too many responsibilities.

* * *

MY WIFE, MY EQUAL

(This article first appeared in *Chatelaine,* September 1985.)

A year ago, while I was working on a TV film in Los Angeles, I called Tara in Vancouver. "How did your day go?" I asked. "Well," she said, "I took the children to The Bay for a couple of hours. I got Severn a winter coat and two dresses for Sarika." "Well, that's great," I said, "but what did you *do*?"

Six months later, Tara called me from Boston, where she commutes from Toronto to teach writing at Harvard. "How did your day go?" she began. "Well, I took the kids to the Eaton Centre, and we had a great time looking at fish in the pet store. Didn't manage to buy them anything, except a stuffed bear." "Good for you," she replied wryly. "But what did you *do?*"

What a difference six months make!

Tara has been a committed feminist as long as I've known her, and has taught me that it's one thing to *talk* a good line about being a liberated male but it's a totally different thing to *live* it. Her incredible patience has helped me slowly to overcome a lifetime of conditioning as the only son with three sisters in a traditional Japanese-Canadian family.

I have tried to be as supportive of Tara as she has been of me — from her determination to keep her maiden name to her decision to go on for a Ph.D. at the Unversity of Wisconsin to her application for a faculty job.

Tara is immensely talented and fills many roles — not only as my partner and friend. She is a wonderful daughter, daughter-in-law, stepmother and scholar. While working on her doctorate, she even managed to have two children. She earned that job at Harvard, but it has turned our lives upside down. I was coming down to the wire in a three-year television series called "A Planet for the Taking," as well as hosting "The Nature of Things." Severn, then aged four, was to be enrolled in a French-immersion kindergarten, and Sarika, at eighteen months, had become much more active. Toronto became our base of operations, instead of Vancouver.

Now, I would have to put my body where my mouth was. We had to have someone — me — to be responsible for the children, while Tara was in Cambridge, Massachusetts, three nights a week.

Now, instead of Tara's twisting her schedule and life to fit mine, I was going to have to conform to the needs of the children and *her* career.

It hasn't been easy. For one thing, as the final

shooting for "A Planet for the Taking" took place, I *had* to be at work, yet during shoots I found myself constantly worrying about Severn and Sarika instead of my lines. Life became one mad round of rushing to complete commitments, then tearing off to attend a birthday party or school skate sale. To be frank, I'm still only playing at it; we have a nanny, which lets me go to work and make short trips. And when Tara comes home, she is so delighted to be with the children that I just lie back and let her take over.

I loved the role of traditional father — playing with the children when I came home from work and taking the family camping. But this is different. Being a single parent is terrible! It is impossible to concentrate fully on work — there is always the dread of the emergency phone call in the middle of the day (as when Severn's school bus was hit by a car). This is familiar to every mother, but I had no idea.

But the most crushing part of the single parent role has been the dozens of little things I do every day *without acknowledgement*: dressing the children, making their beds, picking up and putting things away, doing dishes — a never-ending battle against *entropy*. No one ever says "Thanks," or "Good for you, David." That's all it would take to make me feel great.

I have two concerns about changing sex roles. First, Tara seems able to give herself to her children totally. I have to struggle to do it — I'm often juggling my schedule or a business commitment in my mind, while going through the motions of parenting.

My more serious concern is with the consequences of what Tara and I are doing as we try to define new roles. We are wracked with *guilt* about what we are or are not doing for the children and our own relationship. We love our jobs, yet feel wretched that they divert time away from the family; on the other hand,

> we feel we're not doing the jobs as well as we could because we try to stretch family time. Our friends tell us that our daughters are very lucky to have us as role models, that it's good for them to see their father scrubbing floors and their academic mother commuting to work. They claim that it's *quality,* not *quantity,* of time that matters.
>
> But I know that we *don't* know whether our daughters are lucky. This is all very new. So we muddle along doing the best we can, hoping our life styles don't leave the children psychically scarred. But now my life is settling down, and I've already learned that a day running the house and enjoying the children is quite an achievement.

* * *

As a born-again parent, I look to the future world my children will inherit and what I see impels me to speak out on the approaching global crises. I'm often accused of being a pessimist, but I don't think I am. It is true that I believe if we continue as we are going now, then the global ecosystem and human beings have had it. Species extinction and habitat destruction are going on at a terrifying rate. It is an untenable conceit to believe that we can maintain our current rate of consumption of energy and resources and that the environment will absorb the massive amount of pollutants and debris we dump into it. It is a delusion to think that we know enough to control, manipulate and manage nature. All projections of human activity lead to predictions of energy depletion, massive species extinction and the disappearance of all wilderness by the first years of the next century.

If we don't accept the reality of those simple extrapolations and change our behaviour, then I am pessimistic. What sustains me, though, is an optimism about the strength of love — love of our children and a hope for a better world for them that must surely override all other considerations. And I hope that we will finally see and hear the signs and cries from nature everywhere. That would be the beginning of a wonderful transition in the metamorphosis of our species.

Our children will be twenty-first century inhabitants and we can see only the vague outlines of what their world will be like. But the major reality in their lives will remain nuclear weapons and the arms race. Global military commitments consume a staggering proportion of our efforts and resources. It is estimated that seventeen billion dollars annually will feed the hungry masses of the world. The military around the world spends that amount every ten days. It's a fact that calls into severe question our species' claim to intelligence. Children sometimes see the dilemma more clearly.

* * *

A GRIM FAIRY TALE

(This article first appeared in *The Toronto Star*, May 1986.)

My daughters love the Royal Ontario Museum in Toronto. Fortunately, their dad never tires of it either. Our favourite displays, of course, are the dinosaurs and other prehistoric animals. The other day as we approached the awesome display of *Megaceros giganteus,* the long-extinct Irish elk with its spectacular antlers, Severn marvelled at their size and asked what happened to them. Taking a bit of paleontological liberty, I told her the following story.

Thousands of years ago, one gargantuan bull elk ruled all of the others of his species throughout Europe. He was the most magnificent elk ever known, with an immense set of antlers, a powerful bellowing voice and a majestic muscular body. And as he strode across the vast plains, the ground shook and all the other elks trembled with fear and envy. With his strength and antlers, he was invincible, overwhelming any who dared challenge his power. Those who shared his territory accepted his authority. Each day, he surveyed a different part of the vast expanse of his range.

One day, he ventured further than he had ever gone, beyond his own terrain. But he was confident in his might; after all, his antlers were the greatest weapons ever developed. Presently, he spotted another elk in the distance. "Who could be so impudent to encroach on my turf?" he wondered and hurried toward the figure. He recognized him as a stranger by the colour of his fur. The foreigner was every bit as big and carried a set of antlers as impressive as his own. Trembling with fear and rage, he ran at the stranger and bellowed with all his might, "This is my territory! *I* rule it! Submit, back away or be destroyed!" He was shocked to hear a reply that rang through the air, "No, this is *my* land and *you* are trespassing! *I* am the commander here!" Pawing the ground and shaking their antlers, the two giants stood facing each other, threatening and yelling all day while the smaller elks caught between them cringed in fear of being crushed by these superb animals. When night fell, the two bulls retreated to their respective camps and called for their scientists and engineers.

"I have met a mighty force," each told his respective minions, "and I need reinforcements, an advantage to overcome my enemy's strength. Do everything you can or else he will overwhelm me and subjugate you in slavery." And so a great effort was made. One had many long, sharp spines added to his antlers to stab his opponent, while the other had a large, heavy club placed up front to deliver a knockout blow. The next day the two stags hurried back to the border only to discover the ingenious changes added to each other's antlers overnight. Once again, they stood shaking and roaring but each fearing the other's possible advantage. When night fell, they rushed back for additional reinforcements and defenses against the enemy's clever inventions. Antlers were armoured with flat surfaces to blunt the other's blows and embellishments were added at the edges to probe other weaknesses in the defensive shield.

And so it continued day after day. Each night, the bulls required more help and demanded that those under their control contribute more resources, muscle power and imagination, for victory by the other male brought terrifying possibilities. But each new addition to those giant antlers had other repercussions. The weight of the racks grew so great that the bulls' necks had to be shored up with muscle and bone. But now they couldn't turn their heads as quickly and so they needed other elk to provide an early warning of danger and possibly even absorb the first blow. Legs had to be increased in diameter to support the massive weight and to generate the driving power to wield the antlers. Their bodies were increased in girth to provide more lung power and stomach volume to fuel the muscles. More and more material, effort and creativity went into supporting those magificent antlers.

Inevitably, rumbles of discontent spread from the lower ranks of elk. "Those antlers are draining resources from everything else," they grumbled. "Wouldn't it be better to sit down and discuss a way to co-exist, perhaps share space, and maybe even co-operate?" some suggested. But the two males bellowed at such treachery. "How dare you consider co-existence with a tyrant? The best protection is a superior offensive capacity and an invincible defence. We have to develop cleverer ways to gain an advantage." And when some asked, "Is it worth the expense?", both of the great elks replied, "Of course it is! There will be enormous spin-offs. You'll have more shade from my antlers on hot days and birds will find greater space to perch on." And so it went. Each new development led inevitably to more complex and contrived inventions. Those antlers were an obsession to the two bulls and came to dominate every other elk's life.

"And then what happened, daddy?" asked Sarika, my three-year-old.

"Well, dear," I replied, "eventually the smaller elks realized that the two opponents had put all of their faith in massive antlers that looked impressive but were completely unreliable and impractical. So they simply left the two giants alone to roar and threaten each other. Eventually the sheer weight and cost of those antlers broke their backs and both of them died. The other elks were grateful that they hadn't developed such useless structures — their antlers were quite big enough. So they lived together in herds and turned their attention to the important business of living."

"Oh," said Severn, "and that's why Canada shouldn't get involved in Star Wars?"

"Yes, sweetheart," I said. Even a six-year-old can see the obvious.

* * *